# The Great Barn of Bredon

ITS FIRE AND REBUILDING

*by*

F. W. B. CHARLES

*with a contribution by*

Christopher Dyer

*and original photographs by*

Martin Charles

Oxbow Monograph 76
1997

*Published by*
Oxbow Books, Park End Place, Oxford OX1 1HN



ISBN 1 900188 27 9

This book is available direct from
Oxbow Books, Park End Place, Oxford OX1 1HN
*(Phone: 01865–241249; Fax: 018665–794449)*

*and*

The David Brown Book Company
PO Box 511, Oakville, CT 06779, USA
*(Phone: 860–945–9329; Fax: 860–945–9468)*

Printed in Great Britain at
*The Short Run Press, Exeter*

# CONTENTS


| | | |
|---|---|---|
| Introduction | | v |
| Chapter 1 | Bredon Barn in context | 1 |
| Chapter 2 | Bredon barn in history *by Christopher Dyer* | 23 |
| Chapter 3 | The Fire | 29 |
| Chapter 4 | Emergency repairs | 47 |
| Chapter 5 | Reconstruction | 59 |
| Notes | | 95 |

# INTRODUCTION

Great medieval barns have never received their due. Only William Morris, possibly with a sprinkling of his customary over-zealousness, described Great Coxwell barn as: "one of the finest buildings in England or in the world, unapproachable in its dignity, as beautiful as a cathedral, yet with no ostentation of the builder's art".[1]

Otherwise, Walter Horn in his pioneering studies of timber structures is alone in his appreciation of them. He has analysed their design and construction, showing that barns were not simply functional, or blown-up vernacular architecture, but rather designed by the same master carpenters and masons who were responsible for the great churches and monasteries.

Horn noticed that the design of Great Coxwell's timber-frame had a close resemblance with that of the gigantic barn of Ter Doest, on the Belgian coast near Ostende, and with the one-time frame of the even larger barn of the Abbey of Beaulieu St Leonards in Hampshire.[2] Though nothing was left of its roof or framing, except one or two vestiges of trusses within a later, much smaller barn built within its site, the "shadows" of the end truss could be discerned on the masonry of the one still standing original gable wall. These showed the same system of bracing as that of Great Coxwell and Ter Doest, stamping the three as of the same school of carpentry. All were Cistercian.

A more sobering fact to emerge from Horn's research is that before the Dissolution there were at least 2000 Cistercian barns alone in England. His estimate is based on the knowledge that there were 75 Cistercian abbeys and 26 Cistercian nunneries in England. Beaulieu Abbey, probably of average size, is also known to have possessed 27 outlying manors and granges, each with its barn, built as a matter of priority when each settlement was founded. Great Coxwell was one, and while not all would have been as grand as this, still less as huge as the barn at Beaulieu, the loss of all the rest may seem incomprehensible. The recently restored Abbey Barn at Coggleshall, Essex, was also Cistercian but conformed more with English Lowland practice of the twelfth century than with what we know of Cistercian style. But add to these the barns of the other orders, of private landowners, such as the Bishop of Worcester who built Bredon, and all the powerful lords, ecclesiastical and secular, including the crown itself, and the total of barns destroyed must run into tens of thousands. The wastage, though now a relative trickle, still continues.

When I telephoned Horn in California about the burning of Bredon, he wrote in reply, "of about a dozen medieval barns in timber visited and surveyed by Ernest Born and myself in England in 1960 and 1961, four have since perished in fires: in 1962, the 15th century manor barn of Nettlestead Place, Kent, was completely destroyed by a fire, set by an undetected spark that jumped from a bailing machine into a heap of straw. In 1963, the great 15th century barn of Cumhill Farm in Pilton, Somerset, was struck by lightning that set its roof ablaze and destroyed it to the last timber. It had never been properly surveyed. In 1964, the lesser of two large barns at Lenham, Kent, of probably the beginning of the 14th century burned to the ground after two children threw a burning match into a stack of hay. Last and most distressing .... Bredon".

Bredon alone has been rebuilt. This book is concerned with its story. It begins with the reluctance of the National Trust and ends with a great barn restored so that the differences between before and after are negligible. But that it is substantially a rebuild of 1981 is not concealed. On the contrary, the date is given on one of the corbel stones that support the first part of the frame to be erected, both at the first build and at its reconstruction, and also by an oak plaque naming all those craftsmen and workers employed by the contractor throughout the reconstruction. The National Trust, who of course eventually came round to rebuilding, have also mounted exhibition panels in the main porch of the barn to explain its life history, including the fire.

Also briefly mentioned in chapter 1 are five other barns. This places Bredon in its context of buildings of the same scale and in the same region, and emphasises the uniqueness of its design. Unexpectedly, it is not the only barn to have been rebuilt after fire. The other was Frocester, five hundred years ago, but the circumstances are so different that the two are hardly comparable. Another reason is that of the remaining four, three have been restored, Middle Littleton by us, also for the National Trust, in 1977, and two of them by the State. The restoration of Bradford-on-Avon barn was undertaken by the old Ministry of Public Building and Works in the 1950's, and of Leigh Court barn by its successors, a department of English Heritage, completed in 1987. The differences between our work and that of the State may be a matter of principle, but also, perhaps more decisively, of practicability and cost. The State, as Sir Mortimer Wheeler once remarked, restore their buildings like battleships so that they will never again have to be touched. They must also appear exactly as the original, all earlier repairs eradicated. We, on the other hand, making virtue out of necessity, believe in visible repairs as the more "honest" approach.

Their method can be discovered in the finished building only by the closest scrutiny down to the last peg (even then, much must be guessed) or by being granted the privilege to visit their works in progress, or even by a little subterfuge. I have had to exercise all three to discover the extent of hidden restoration of both of the barns, particularly the concrete foundations at Leigh Court that would sink a battleship.

There are six or seven other medieval barns within our region that have not been mentioned. They can add little to the chosen six, either in architectural design, their size or their state of preservation. All should be written up (if not already) and fully surveyed, so that should any of the disasters habitual to barns recur, they too may be rebuilt, or at the very least adequately recorded. The great barns outside our region – Harmondsworth, the two at Cressing Temple, Coggleshall, and others, wholly timber-framed and therefore still more vulnerable than Bredon – are also omitted. But these were already written up and fully recorded.[3]

The most serious current loss, or near loss, is the longest of all the West Country barns, that of Abbotsbury (276 feet) near Dorchester. It is also architecturally the richest, not even excluding Great Coxwell. But now only twelve of its twenty-four bays are roofed. The unroofed walls are growing ivy and gradually crumbling. As most great barns that have been in trouble if not already gone, this barn is privately owned, and there are few private owners, with the notable exception of Mr Edward Price of Frocester, who have either the means or devotion to maintain them.

*Acknowledgements*

Any virtue this book may have is due, first, to Walter Horn who invited me into his team as architect when his first choice, Ernest Born, dropped out. He was the guiding force behind our joint surveys of Middleton Littleton, Leigh Court and Frocester, in that order. Bredon was next on our list.

We had examined it while our dendrochronologists and carbon-14 experts were taking samples of timber for dating. We then discovered its construction, its structural details and method of erection, so that measured dimensions were all we needed before drawing-out and publication. Needless to say, without that knowledge, Bredon could not have been rebuilt.

Secondly, the idea of writing a book would not have arisen without the photographs taken from time to time throughout the contract by my son, Martin Charles. He had already photographed for publication some of our earlier restorations. But for Bredon we decided that a fuller record should be kept so that the barn could be seen in its successive stages, starting with the wreck after the fire and ending with the same views after completion. In between were eight visits by Martin at roughly 3–4 monthly intervals. A few snaps of my own fill in the gaps.

Thirdly, I must acknowledge Russell Jones, our office architect for Bredon, whose drawings consist of three full sets, the first of all the trusses and longitudinal framing as surveyed immediately after the fire, accompanied by proposals for repairs or replacements, secondly a more accurate, larger-scale set of S and R drawings, made when the trusses and frames could be more closely examined on the floor or bench, and thirdly the set of "as-mades", showing the repairs as actually carried out. Russell was also regularly, at least one day a week, at the barn, drawing details on the spot and constantly advising.

Bracketed with Russell Jones is Gerald Dauncey, the foreman carpenter of Spicers of Worcester, the contractor. Such a contract is by its nature cooperative – or disastrous. Although there was no certainty during the first six months about rebuilding, Spicers, in the person of their managing director, John Chambers, kept their men on the job, in particular Gerald Dauncey, so that he knew as much about the building and the mysteries of medieval carpentry as even Russell Jones when the time came for putting it all together again. Also linked with the building team, and not for the first time, was Paddy Firminger, the Quantity Surveyor. Without him and his office I could not have kept the job going, nor have controlled it when it properly started. A surveyor of his calibre means much more to the architect than the one who merely keeps accounts and prepares estimates. Most important, he had already gained the respect and confidence of the National Trust on our earlier contract, the restoration of Middle Littleton, so that the regional office knew that they would not lose control of finances, despite the frequently unconfirmed instructions I had to issue.

Alastair Finlinson, the regional director, who for us was the National Trust itself, also has my special regard and gratitude.* No building owner except the Trust would have rebuilt Bredon. For many reasons it is certain that the State would not have done so, even if the barn had been in state care. First, Bredon was only a barn, not a royal palace, and, according to Stuart Rigold, the top historian and adviser of the Ancient Monuments Department, grants could be made only for preservation, not for restoration and reconstruction. Again, having seen the State's approach to restoration, it is no less certain that there would have been very little of the structure still standing by the time they had got round to preparing a scheme. Nor had they, nor would they have accepted, first-hand information such as ours. For when the state takes over, nothing can be left to others. In brief, reconstruction would have been impracticable, regardless of cost.

By contrast, Finlinson never doubted that Bredon had to be rebuilt. On that black Friday night as we and many others, several weeping, watched the blaze and the firemen fighting it, Finlinson asked how much it was going to cost. I guessed, as no architect should ever do, a quarter-of-a-million. And, of course, that figure stuck. Six months later, the Trust and Spicers signed the contract for £251,254.00.

In 1984, Christopher Dyer, Professor of Medieval History at Birmingham University, invited me to share with him a half day course at Bredon under the title "The Bredon Medieval Barn : its History and Architecture". Returning the compliment, I invited him to put down in writing his contribution to that course. Chapter 2, is the result adding another dimension to this book and so to appreciation of the barn. It is also a privilege for me to have our names once again linked.

My last and deepest acknowledgement is to my wife, Mary. Her work is visible throughout in the form of immaculate drawings and invisible, but still there, in every written page.

## *Technical Preface*

The carpenter's yard was the focus of all medieval framed structures. Occasionally the site of the future building could also be the yard, but for many reasons this may be discounted at Bredon. It was already the site of a large and wealthy barton with the bishop's manor house close by. There was probably already a barn there, perhaps timber-framed and requiring replacement with something bigger and better, if not more defensible. But the most telling reason is that the clear methodical assembly numbers scribed on its timbers show that every frame had been prefabricated and its components delivered to the site ready for erection.

The perfection and uniformity of its joints also speak of its framing under organised conditions by carpenters who not only knew every trick of the trade but whose craft was lifelong and shared by all from the master carpenter downwards, the master carpenter and master mason being the architects responsible for the design of the building, as instructed no doubt by the bishop himself, and for its entire construction.

In Chapter 2 we may see how the project was probably started, at a meeting chaired by the bishop at his palace at Hartlebury or Worcester with his chief officials. One may hear the bishop's advice to his architects: "I want none of these "cruck" affairs, as Evesham and Pershore still go in for, but a permanent structure with proper pillars or posts to carry the roof, as they build and always have done on the continent, or even in our own country where bishops aren't so poor as we are! And you will use our wood at Welland. And you will not make a great mess at our farm in Bredon. And, yes, my dear Peter (see Chapter 2) there will be a proper reeve's office, where he can see that none of those miserable peasants steals or cheats us – day or night..."

So the barn had to be carefully planned and designed by the architects answerable directly to the bishop or his steward, the stone ordered (the barn shows that some of the stonework may have been salvaged from an earlier building or

---

*I am also grateful for his reading the text of this book, especially since he found only two errors of fact. He was not at Bredon on the night of the fire. Our conversation, reported in the next paragraph, took place therefore on one of the following evenings when we and others were still looking at the smouldering wreck afraid that it might flare up again and destroy what was left. Secondly, on page 54 I have credited him as present at the important site meeting on 28 July 1980. He says only his deputy was there.

buildings), and the trees obtained from the bishop's woodlands at Welland. They were selected for straightness, or uniformity if intended for the many curved or "swept" members, and the choice was always for the smallest tree that could yield the required finished dimensions.

Probably the swept components came from coppiced oaks which would grow naturally with a slight bend. Felled, debarked and sawn to length, on the edge of the forest, and stacked according to their future functions, posts, purlins and plates, would be squared by axing, beams and braces 'halved' by rift-sawing on a frame or over a saw-pit, and lesser members, such as rafters and struts, put aside for future needs.

This was 'first conversion'. 'Second conversion' was the laying out to string-lines on a level floor or prepared ground each cross-frame. The beams and braces overlapped the posts and were accurately marked for jointing. At the same time each component was numbered. The 'upper' face was the halved face or, for the posts, the uppermost face as laid out. All components to left of the centre-line of each truss or frame had plain Romans: I, II, III, IIII, etc. Those to the right were flecked: λ, Iλ, IIλ, IIIλ.... Following the cross-frames I to X, the arcade-framing was laid out, bay by bay, and again the joints, including joints to the sides of the posts, marked on the timbers. Numbering was the same except that the arcade-plate of bay 1 was numbered I and II, as well as the braces. So the numbers through the length of the barn were doubled up and instead of I to X as the cross-frames, the longitudinal members started at I and finished at XVIII, that is, two numbers in each of the nine bays.

Incidentally, the reason for the duplication up to four times of the digit I was to prevent IV from being mistaken for VI, and the inversion of the V was to prevent its becoming X should the gouge, drawn towards the body, slip beyond the meeting of the strokes.

Second conversion, including cutting all the joints, was of course by far the longest of the series of operations. It was also the operation, because it took place remote from the site, that was invariably unnoticed in documents. So one reads of a building being raised in a week, or a large building contract completed in a few months, when, in reality, from first design to opening day the period was probably a matter of years.

The only major works after second conversion were transporting each disassembled frame to the site. For Bredon the frames were probably carted to Upton, floated down the Severn to Tewkesbury and from there again carted to Bredon. This was probably easier than towing them upstream on the Avon and dragging them up the very steep bank to the barn site.

They then had to be laid out again in their exact positions for raising. Boring and pegging was the final conversion. Then the drama of rearing each truss and erecting the bay framing on either side so that each successive frame was secured, until the whole of the main framing was up. Then back to the beginning again for the upper roof, followed by raftering and finally the roof-covering. Thus, as the photographs of the rebuilding show, the construction process resembled a series of waves coming in with the tide from south to north.

# Chapter 1

# BREDON BARN IN CONTEXT

There are six great medieval barns still standing in the Cotswold-Severn region from Bradford-on-Avon in the south to Worcester in the north. They are Bradford-on-Avon, Frocester, Great Coxwell, Bredon, Middle Littleton and Leigh Court.

Worcestershire is exceptional in possessing three of them – Bredon, Middle Littleton and Leigh Court – all within twenty miles of Worcester and about the same distance from Tewkesbury. They represent the three main forms of timber-frame construction of the middle ages – aisled, cruck and base-cruck.

Others might be added to this list, but these are the largest, best preserved and oldest. Together they present an astonishing range of structure in their timber-framing, their overall dimensions and architectural design. There is not one that could be described as anything but unique and irreplaceable. Four of them are of cruck construction, and since they are all within a major cruck region extending from North Wales to the English south coast, it may be wondered why two of them, Great Coxwell and Bredon, are aisled. Yet so different are both of these from each other and from the aisled halls and barns of the home counties and lowland England that reasons other than regional influence must be sought to account for their design.

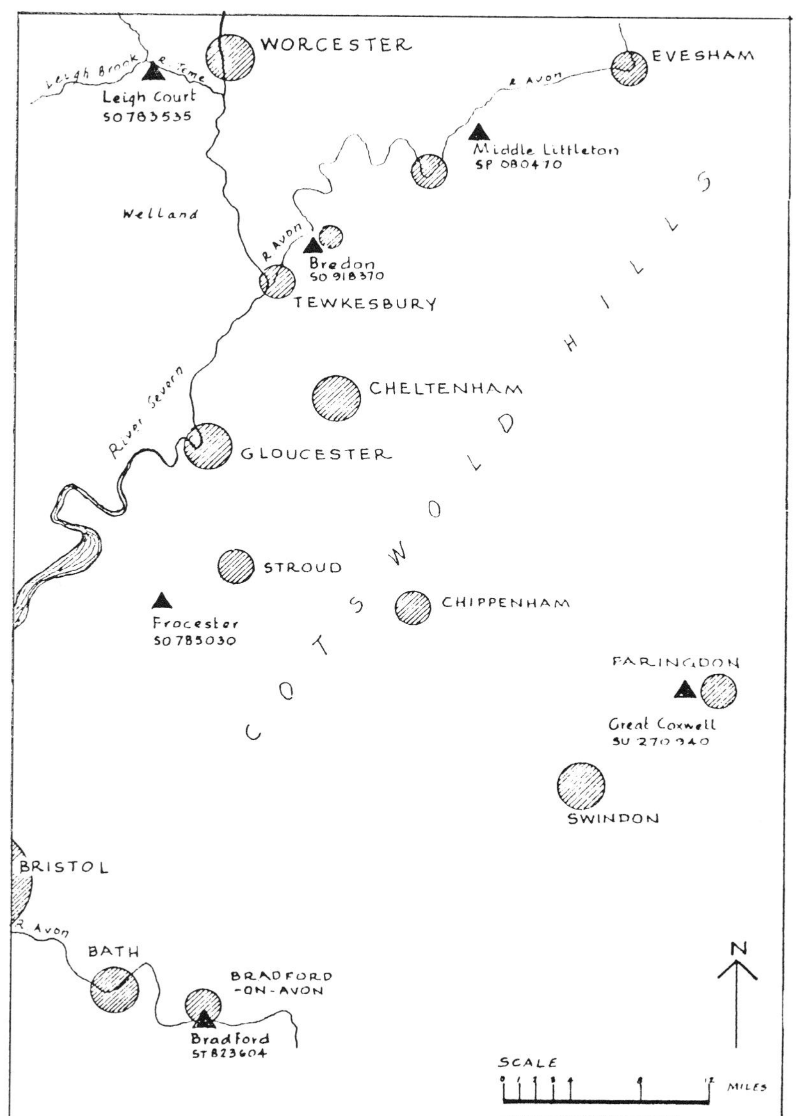


*Fig. 1. Location map of the six great barns of the Severn – Cotswolds region, giving O.S. numbers*

## *Leigh Court*[1]

First, Leigh Court is the only barn that is not stone-walled and Cotswold slate-roofed. Instead, it is wholly timber-framed standing on a four foot high local red sandstone plinth wall. The roof is clay-tiled, possibly thatched originally. It is the largest simple-cruck structure in England and probably anywhere. There are nine trusses or cross frames that reach from the plinth wall to the roof ridge, or very near it. The two end trusses are "base-crucks", cut off half-way up the roof and connected by the collar beam to form a half-hip at each end. The blades of the main trusses, of which there are eighteen, are between thirty-four and forty feet long, slightly bent or swept, some double-swept. They are too large to have been obtained by the method of halving, so producing mirror-images. Instead each blade is the stem of a pretty large oak tree, selected so that it could be paired with another of practically identical shape to form a symmetrical truss. All the trees were local, yet today it would require a search far and wide to find one to replace any of the blades, let alone a matching pair.

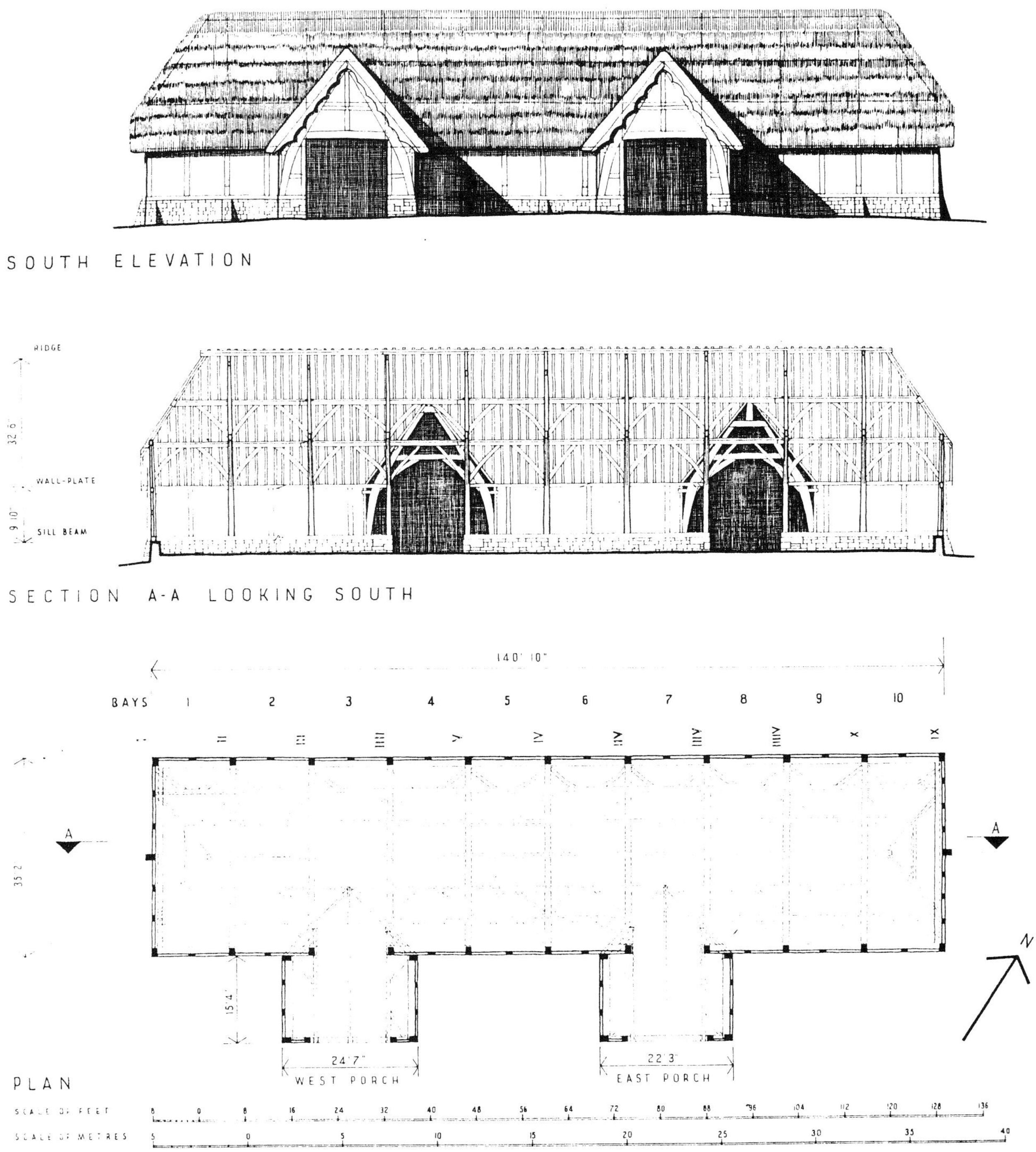


*Fig. 2. Leigh Court barn – south elevation, section and plan*

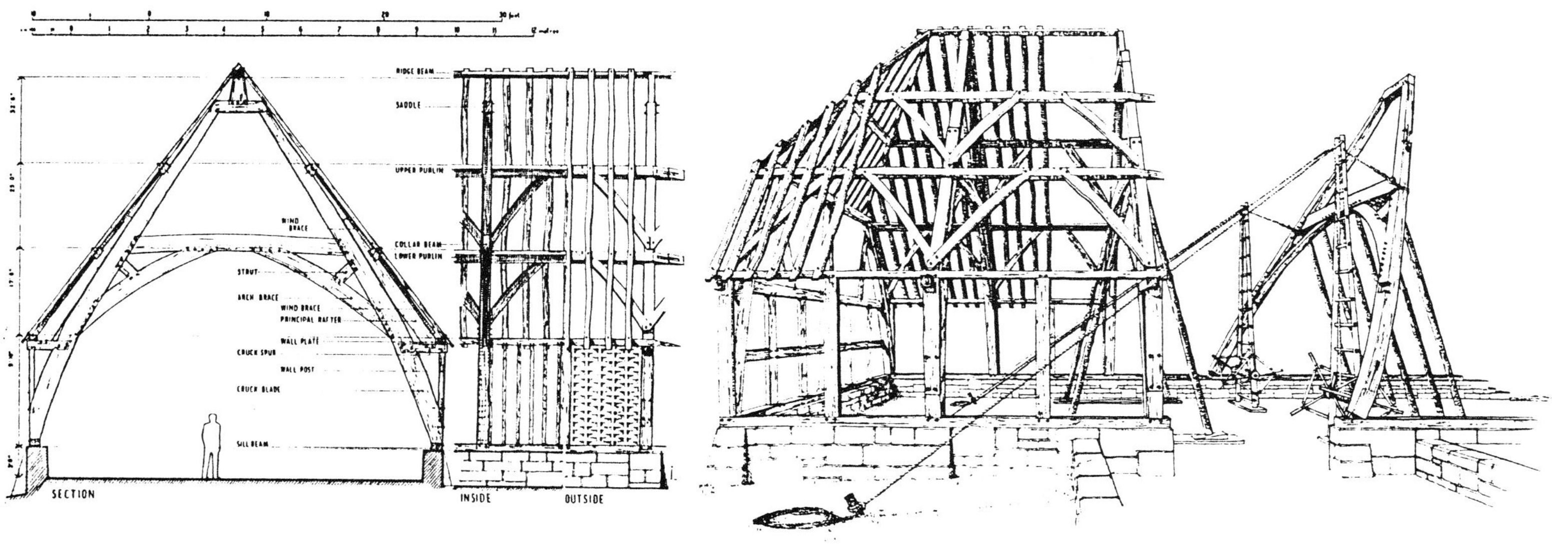


*Fig. 3. Leigh Court barn – section and rearing truss IIII*

*Figs 4 & 5. Leigh Court barn – interior looking west and exterior from east (photos: Horn)*

Dendrochronological tests at the time of our survey showed that the oaks were uniformly of rapid growth, their annual rings very wide and "complacent", and therefore useless for dating. But carbon-14 analysis produced a fairly reliable date of the beginning of the fourteenth century.

The method of erection was clear, not from the numbering of the trusses, as the numerals were cut on the shoulders of the joints and so concealed except where joints had opened. Instead, the upper and lower face of each truss was obvious, showing that they were reared in sequence from the west gable to the east. A refinement of the rearing process was discovered during the building's restoration in 1989. The purlins span two bays longitudinally. They could be placed therefore when the first two trusses had been reared and then cantilevered and propped across the next bay, awaiting the third truss. Trusses II, IIII, VI and so on down to X would thus have to be temporarily stabilised after rearing. How this was done was shown by the existence of a pair of holes near the ends of each collar beam. One of each pair had been drilled at 45° downward in one direction, the other downward in the opposite direction. Thus four ropes, two in each direction, were used as stays. There were no such holes in the trusses with odd numbers, since these would be secured upright by the awaiting purlins as soon as they were reared.

Though cruck construction is often looked on as primitive, there is nothing of that nature in Leigh Court. As well as the largest it is probably also one of the most sophisticated cruck buildings in existence. The richness of its design, notably the carved bargeboards of the porches, which are also cruck-built, its exact and intricate carpentry and the scale of the building distinguish it from most other crucks. It was owned by Pershore Abbey.

It is not difficult to account for the design of Leigh Court. In the woody area of the Teme River, just before its confluence with the Severn, oak trees were evidently large and plentiful, and could be felled and converted even while erection was under way. There was probably therefore no separate carpenter's yard, the entire building process, from selection of trees to final erection, taking place on or near the site.

At the time of our survey, in 1967 and 1968, the structural condition of Leigh Court was not serious. Slipped tiles were causing leaks that could threaten the more or less irreplaceable cruck blades but they were not yet affected. Spread of the trusses was apparent and every truss had been strapped between the collar beam and arch brace on its south side. Further strapping would probably have been adequate to stop further movement. It speaks volumes for the strength of the frames, particularly their jointing, that so little had had to be done in nearly six hundred years. The slight tilting of the trusses on each side of the inserted doorways on the north side of the barn owing to the damage done to the plinth wall by farm machinery was probably the most serious defect.

Our report called attention to all this, but the owner had not the financial means (though only a few thousand pounds at the time) to carry out repairs. Eventually, the Department of the Environment having spent two years deciding whether to take this barn or Middle Littleton into State Care, plumped for Leigh Court. It then needed another ten years to negotiate an agreement with the owner, until at last in 1987, twenty years after our survey, the Historic Monuments and Buildings Commission (English Heritage), with the architect, Michael Peach, previously a member of our office, began restoration.

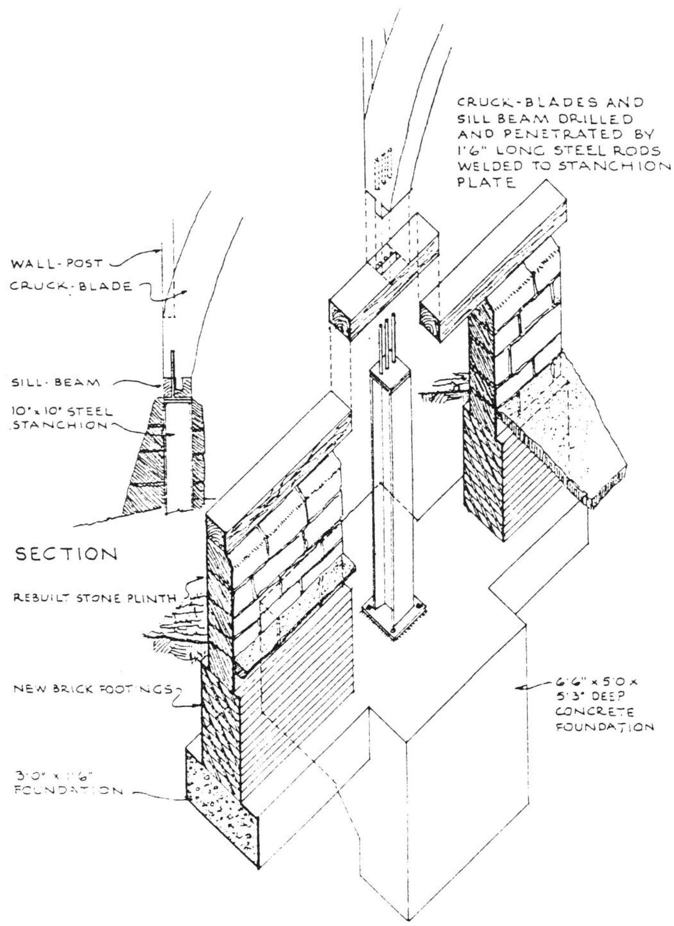


*Fig. 6. Leigh Court barn – axonometric showing foundations by English Heritage*

The scheme designed by English Heritage involved the underpinning of the entire barn and laying massive foundations from which steel stanchions were erected, each with a top plate to which eighteen inch long stainless steel bars were welded. These penetrate the sill beam and cruck feet so that any tendency to spread is transferred to the foundations, restrained against overturning by a huge weight of concrete on the inside, that is, beneath the floor of the barn. The stanchions are concealed within the rebuilt buttresses of the plinth wall.

Thus the trusses retain their form without any more straps or plates, but the cost would have been prohibitive for anybody except the State. Only time will tell the effectiveness of the method. For the present, the barn is better than new.

## *Middle Littleton*[2]

There could be no greater contrast than that between the barns of Leigh Court and Middle Littleton. Middle Littleton is one of our four Cotswold buildings, standing just below the western escarpment of the oolitic limestone belt, on the blue lias. Its walls are therefore lias, the superior oolite from the Cotswold quarries being confined to buttresses, copings, dressings and the remarkable Viking-like finials. It originally had four porches, the large wagon porches on the south side and lesser ones opposite. Both of the eastern porches were demolished, probably at the end of the nineteenth century.

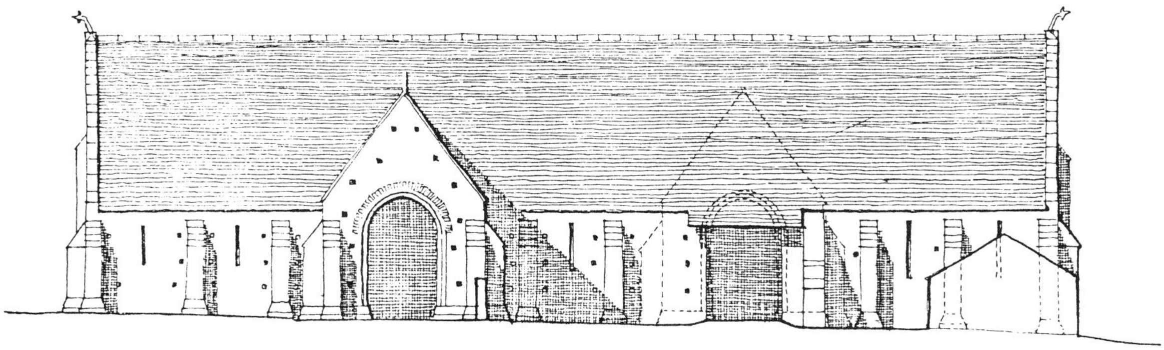

*South elevation*

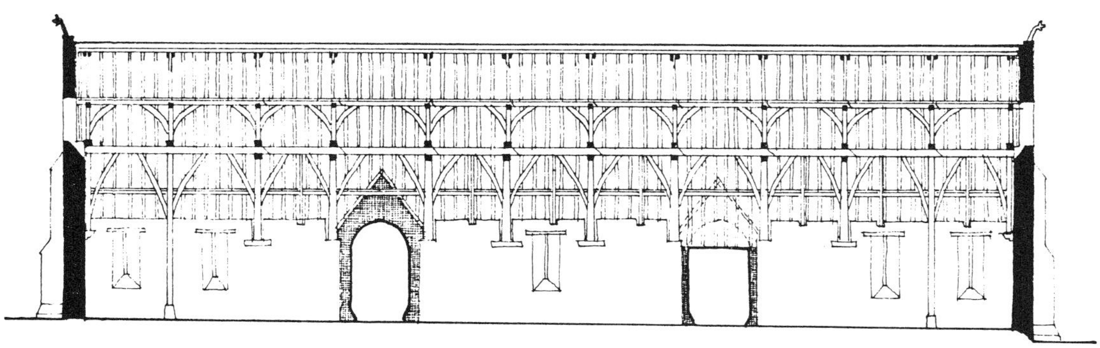

*Long section*

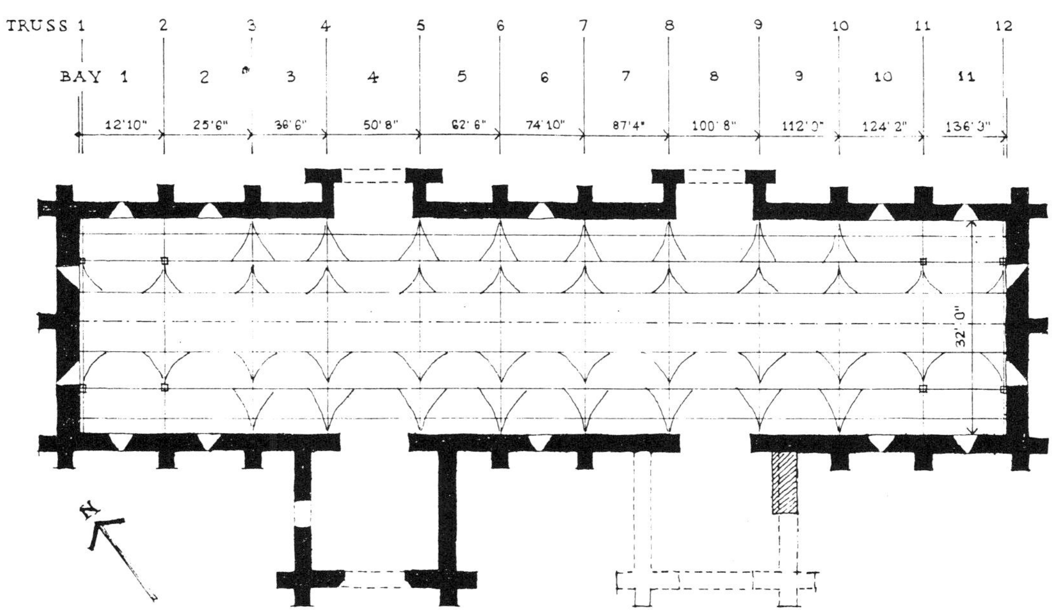


*Plan*

*Fig. 7. Middle Littleton barn*

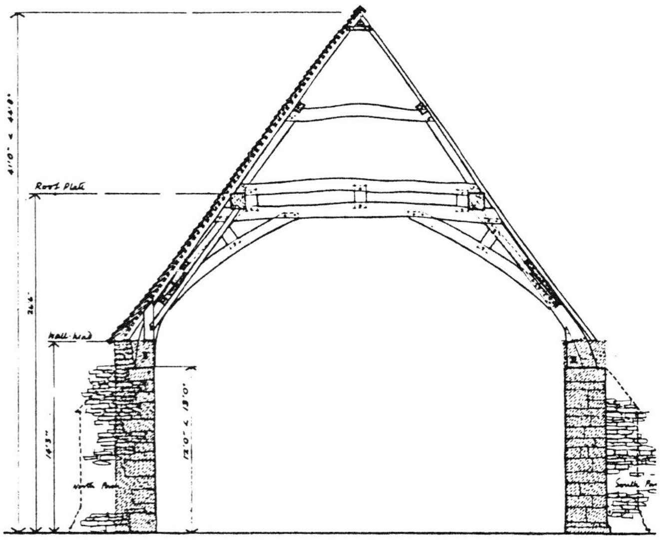

*Cross-section – typical truss*

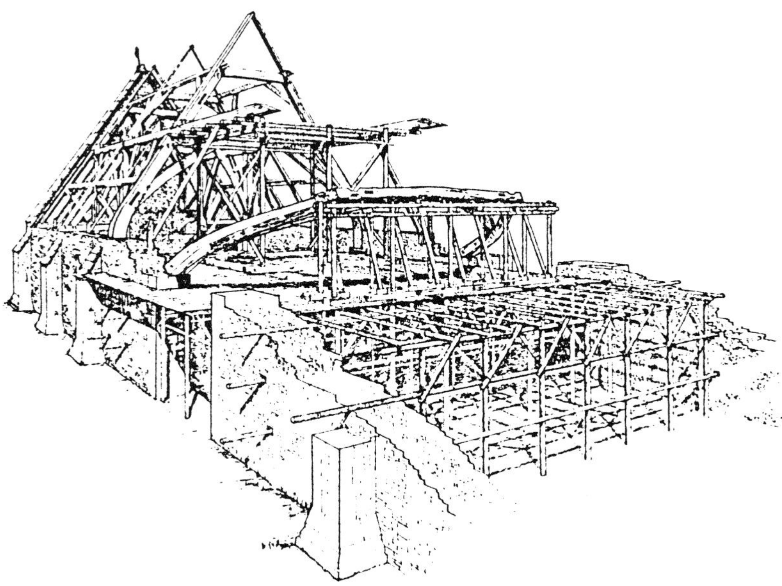

*Rearing truss III (3)*

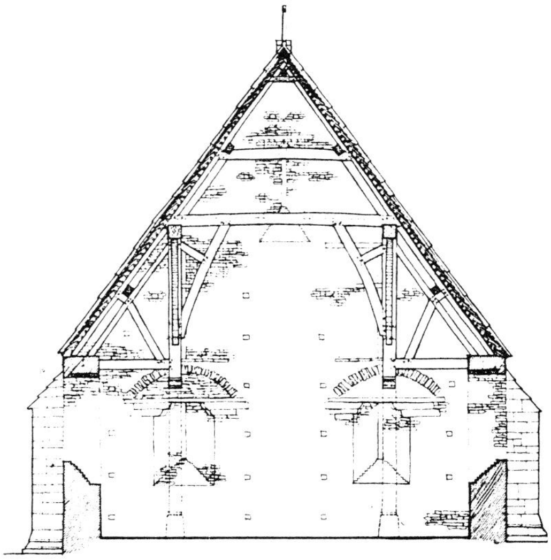

*Cross-section – truss I as built against gable wall, and truss II (2) with full-length posts*
*Fig. 8. Middle Littleton barn*

It is a raised base-cruck structure of eleven bays of which the first and last are aisled. During erection the aisle posts of truss II supported the arcade-plates from truss I cantilevered over bay 2 while awaiting the rearing of the first of the crucks, truss III. The aisled bay at the other end is not simply for the sake of symmetry, but to allow space for setting out and rearing truss X. The barn's design was obviously the mason's, rather than the carpenter's. For the windows of the gable elevations are symmetrically composed entirely for the sake of their external appearance, without thought for the internal timber-frame. As a result their splayed jambs and cills are confused with timbers crossing them, and the heavy corbel-stones supporting the end posts had to be inserted between the wooden lintols of the lower windows and their relieving arches. Moreover, having thus erred at the upper end (the masons had probably had to go back and insert the corbels after that wall had been built) the master mason did not even bother to alter his design for the lower end.

It is however in the quality of timbers that Middle Littleton presents the greatest contrast, not only with Leigh Court, but also with the rest of the barns. The Vale of Evesham was already a fruit (and wine) growing area, in which oak trees were small and scarce. Thus, most of the timbers in the barn are waney edged and irregular in shape, many with the bark still on. But, of most significance, the four aisle-posts, hardly twenty-four feet high had to be made up each in two lengths.

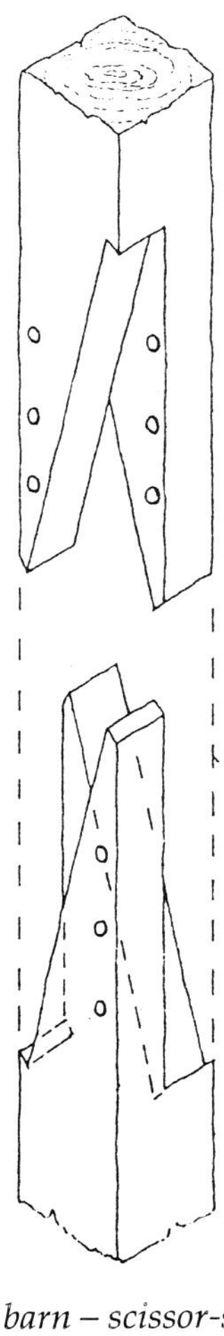

*Fig. 9. Middle Littleton barn – scissor-scarf as found in all four posts of truses II (2) and XI (11)*

Such obvious absence of good timber does not mean lack of carpentry skill. The elegant scissor-scarf of the jointed posts, first discovered at this building and now used extensively for repairs, is itself evidence of high craftsmanship. In addition, accuracy in setting out and erecting so that all joints are dead in line from one end of the building to the other and in exactly the same vertical plane requires more, not less, skill than when the timbers are perfect. Lastly, the numbering of the trusses bears little relation to the sequence of rearing them. There are also two distinct styles of numerals, suggesting that the trusses were prefabricated in two separate yards to be erected as and when they were delivered to the building site.

Middle Littleton, owned by Evesham Abbey may be securely dated to the abbacy of John de Brokehampton, 1282 to 1316, who "built a grange at Littleton", also described in the documents as, "a very fine one".[3] Its condition at the time of our survey in 1965 was considerably worse than that of Leigh Court. The roof was leaking, not through lack of care (replacing slipped stone-slates was a routine occupation of the owner), but because the timber structure was becoming progressively less stable since the two eastern porches in opposite walls had been demolished, about a hundred years ago. The introduction of additional posts and beams had done little to prevent spread and lurching of the roof. The north wall was almost a foot out of plumb at its worst point. After the Department of Environment's decision to take over Leigh Court, a campaign for Middle Littleton had to be started. Eventually the National Trust was persuaded by Christopher Cadbury with pressure from the Society for the Protection of Rural England, to acquire it. After further debate we were appointed for its restoration. The scheme, again in contrast with Leigh Court, was minimal. The most expensive of the repairs was the replacement of four decayed cruck-blades between the south porches. Trees of the right shape could not be found in any timber merchant's stock, but two huge bent butts, too big for conversion by the band saw or for any other purpose, had been abandoned to decay among the nettles in a corner of the yard. These were at last discovered and delivered to the site to be sawn by hand and axed and adzed into the four blades. For the rest, further spread was resisted by tying the cruck feet across the barn with half-inch stainless steel rods. The masonry of the building was scarcely touched, but the contractor in his effort to save money had taken on an incompetent roofer whose first disaster in stripping the first two bays as specified was to throw the precious slates to the ground. Despite error after error, perhaps individually acceptable on the promise that it would be put right, but in aggregate catastrophic, I at last had to condemn the whole roof. Stripping and relaying then had to begin all over again, this time by a skilled roofer. The barn was completed in 1978, ten years before Leigh Court, at the cost of about £53,000.

*Fig. 10. Middle Littleton barn – interior views before and after restoration (photos: M. Charles)*

## Frocester[4]

Frocester, though also a raised base-cruck, as Middle Littleton, is so different architecturally and structurally, and in every carpentry detail, that it required a closer survey than did any of the other barns to discover how it had been built.

First, however, its inordinate length of 190 feet, compared with around 140 feet of the other two, and its relative narrowness, less than 30 feet internally, can be explained by the tendency of barns to become longer and narrower the further into the west country they are situated – another indication that large oak trees were at a premium. Frocester has thirteen bays as compared with the ten and eleven of Leigh and Middle Littleton, respectively.

The cruck-trusses of Frocester however present a number of anomalies: first, the numbering of the trusses is in sequence from the east gable, truss I, to the west, truss XIIII, but the upper face on which the numbers are scribed indicates that erection, if by rearing, took place in the other direction – west to east. Secondly, the pegs were correctly inserted on the same face as the numbers except at truss XIIII. Here they were opposite, a situation that made sense if they had been driven in *after* the truss had been erected, for the proximity of the gable wall precluded the use of a hammer on that side. Thus Truss XIIII had probably been erected piece by piece, as rearing such a frame before it had been securely pegged would have been quite impracticable.

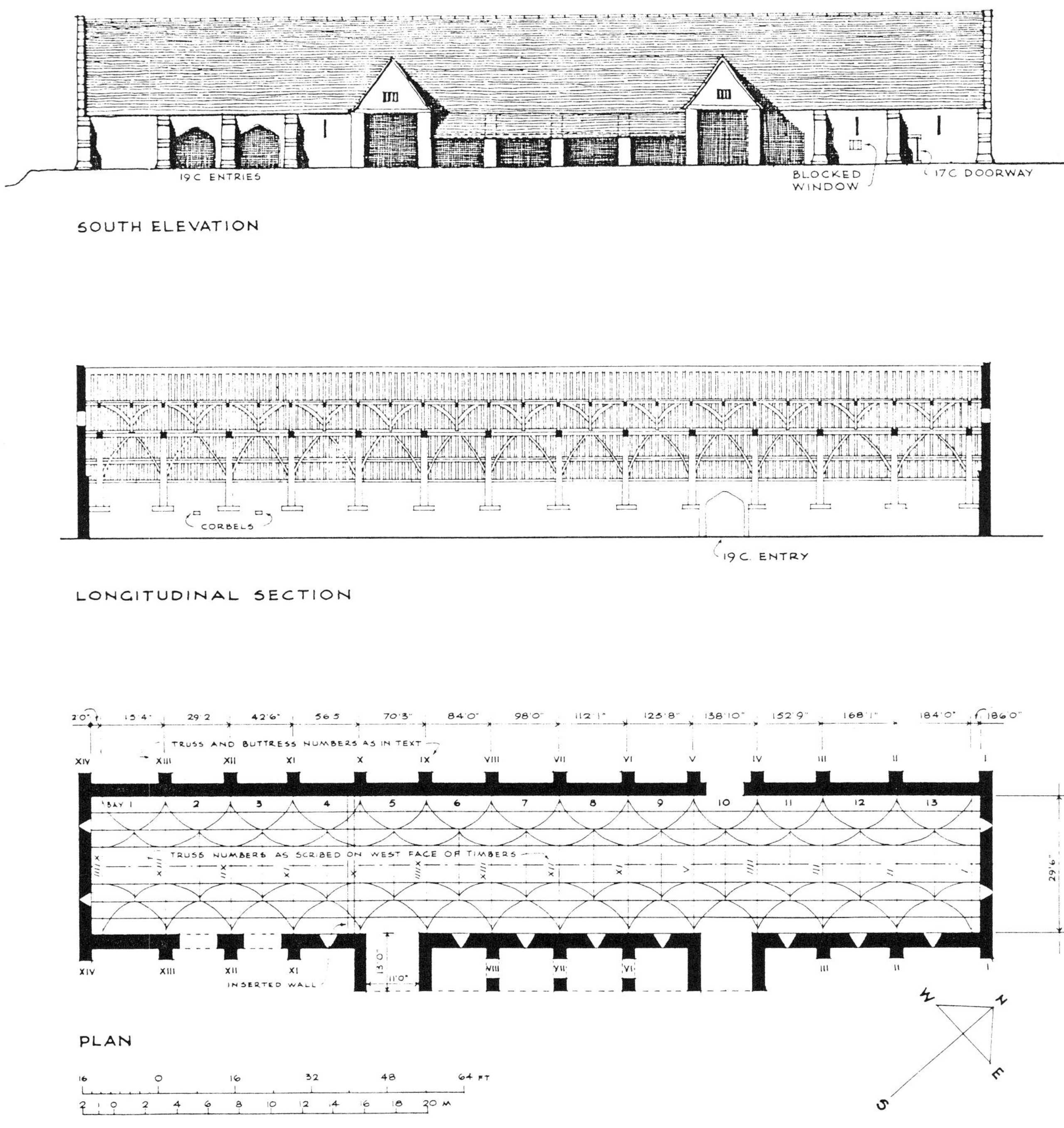


*Fig. 11a. Frocester barn – south elevation, longitudinal section,and plan*

What of the other trusses? One or two had also been pegged on the wrong face, but, most significantly, the joints of the braces to the crucks and to the collar beams of all the trusses had been tightened by wedging. This would not have been necessary had they been framed on the floor or a platform before rearing. In fact it was proof that they could not have been reared.

A final anomaly was that while the main purlins were jointed over the trusses indicating that they had been laid in sequence from west to east, the wall-plate joints showed that they had been laid from east to west. Had the trusses been reared each of these members would have followed the direction of rearing.

So here is a unique example of a cruck structure erected piece by piece, an extremely difficult task, and only necessary if the walls of the building had already been built. The stonework around the cruck feet had indeed been interfered with as if to insert the blades.

All of this raised the question when and why was the roof framed and erected in this way. Like Middle Littleton there was a wholly reliable documentary date for Frocester. It had been built by John de Gamages, Abbot of Gloucester 1284 to 1306, an almost exact contemporary of John de Brokehampton of Evesham Abbey. So the two buildings were probably under construction at the same time. But neither carbon-14 nor dendrochronology produced results to confirm such a date for Frocester. Ten years later, however, our dendro samples were re-analysed under different laboratory conditions and produced the date c. 1524. If that was the date of the present roof what other evidence of the barn's history could be found?

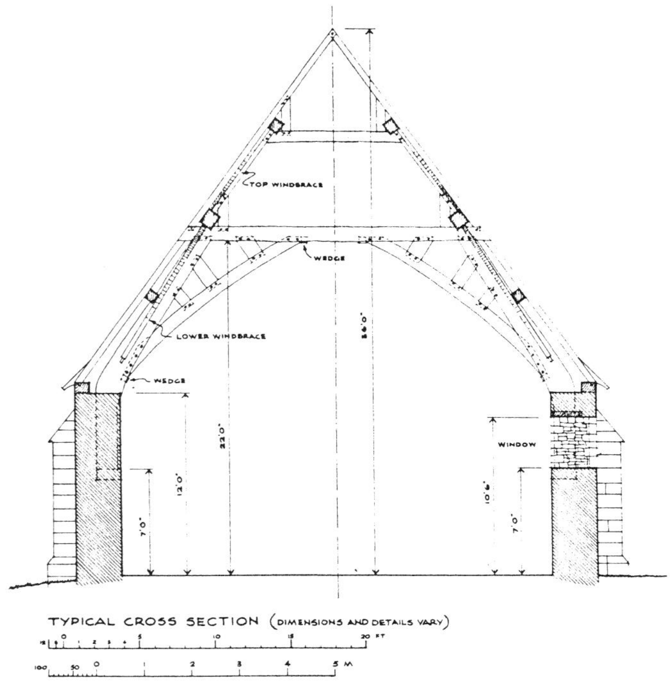


*Fig. 11b. Frocester barn – cross-section*

There were signs in the interior masonry that there had been a fire. We also noticed that where the south wall had been pushed out as a result of roof spread this was not reflected in any deformation of the present trusses. Some of the stonework, notably of one of the buttresses, had been rebuilt vertically, obviously replacing older masonry that was leaning out. Thus, it was the original, medieval roof that had caused these repairs, not the present roof. And, of greatest importance, the roof, unlike the stonework, showed no signs of fire. So it had been entirely renewed, leaving the final question – is it a replica or near replica of the original?

*Figs 12 & 13. Frocester barn – detail showing wedged tenons (photo: Horn) and interior perspective*

*Figs 14 & 15. Frocester barn – cruck foot in rebuilt stonwork and rebuilt stonework in south-east porch*

This may be argued both from the general and the particular. First, it is often wondered why following a disaster such as fire, war or earthquake, rebuilding invariably follows what was there before. In towns of course owners of buildings want back what they had. But also it is more economical to use all original materials, particularly foundations and parts of walls, that have survived. In other words, replication is the easiest, if not the only method. On that argument the present cruck trusses are as originally, conforming with the regional style as at Leigh Court and Middle Littleton.

To reinforce the argument of following precedent: by the sixteenth century cruck construction, if not entirely superseded, was no longer used for large or important buildings. Technical improvements in converting timber, means of transport and changes in economy had tended towards lighter forms. There is some evidence of this in the relative slenderness of most of Frocester's roof components in comparison with the other barns. In particular the reversal of the upper tier of windbraces, springing from the intermediate principals is more "mannerist" than structural and gives a lightness of touch that is not characteristic of medieval design*. Otherwise there is no innovative construction by which the heavy, unwieldy crucks might have been dispensed with and a lighter form of roof built off the wall-head.

So the present roof is clearly and logically a compromise, medieval in form, contradictory in construction and erection method and with minor concessions to up-to-date design. Probably it was considered by its owner – still the Abbey of Gloucester – an important enough building to restore it practically as it had been before the fire.

Its decline began with the Dissolution and its new lay owner, the Duke of Somerset, soon to be bought out by George Huntley in 1554. We can be sure that if the rebuilding of the roof had been done by him it would have been quite different and more utilitarian. He inserted upper floors at both ends of the barn, formed new doors and windows in the east gable wall and was soon to rebuild completely the west gable wall, probably as a result of the disturbance to the original. Incidentally it had been built over a Roman ditch excavated by the barn's present owner, Edward Price. If he was disappointed by the discovery that the roof was not medieval, he was also gratified to own a barn with such a history, now matched only by Bredon.

---

* The wind-bracing at the Brethrens' Hall, St Cross Hospital, Winchester is similar. The Hall is dated 14th Century, but the roof may not be of that date.

## *Bradford-on-Avon*[5]

This barn must be included as typical of the proportions and the raised cruck construction (simple or compound) of barns of the south-west region. It is of fourteen bays as compared with Frocester's thirteen, but nearly twenty feet shorter. The width is the same; the dimensions are 167'6" x 30'6" externally.

It was built by the Abbey of Shaftesbury in the first half of the fourteenth century and is part of the Barton Grange which still largely exists.

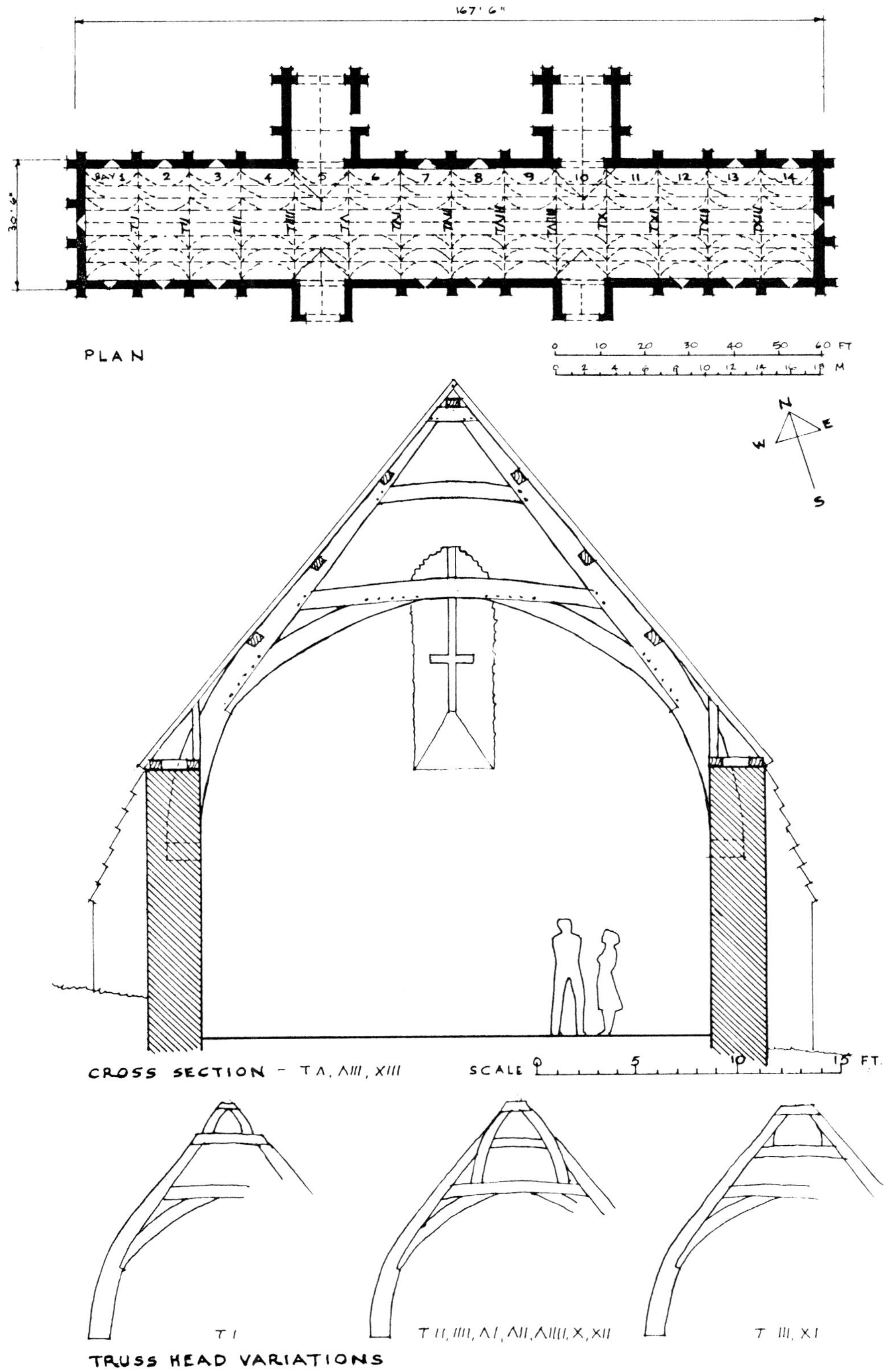


*Fig. 16. Bradford-on-Avon barn – plan, cross-section and the various designs of truss heads*

The site slopes downwards from south to north with the River Avon flowing west and south-west, coming nearest to the barn at its north-west corner. The main front with its two wagon porches faces north. The rear of the barn with its lesser porches is on more steeply rising ground southwards. There is a difference in height between the two walls of at least two feet. The ground also slopes from east to west so that the west gable is considerably taller than the east.

The stone is the most beautiful of the Cotswold limestones, probably from the disused quarry at the top of the hill, of which most of the town is built. The whole of the masonry is in coursed ashlar. The porch doorways have moulded jambs with shallow segmental arches; the buttresses are single stage but with seven weathered courses; and the gable finials originally highly ornate are now seriously worn and depleted, but the main front remains extremely imposing. The proportions of the west gable are accentuated by the twelve foot high loophole with its horizontal cross much nearer the bottom than the top. The whole gable has a primitive verticality, not unlike Bradford's famous Saxon church.

Altogether this barn is more architecturally stylistic than any of the others. Internally the timberwork is also elaborate. There are ashlar pieces to the rafters, seven to each bay with inner and outer wall-plates, a feature shared only with Great Coxwell, and imparting a much more finished appearance to the whole roof structure than is normal.

*Fig. 17. Bradford-on-Avon barn – trusses III and IIII*

*Fig. 18. Bradford-on-Avon barn – looking into south-east porch from main barn showing lower part of trusses VIIII and X and new pads*

But the trusses are the most significant. All are seated on oak pads bedded in the masonry five or so feet below wall head, exactly as the base-crucks of the other barns. And all present the same profile up to main collar level with the heavy arch braces typical of all cruck structures. But above the collar there are four variations in the design of the trusses. Seven of them could be termed two-tier crucks as the blades are tapered off immediately above the collar and abutt the outside edge of a pair of lesser crucks standing on (no doubt tenoned into) the collar. These upper crucks are saddled at the ridge and also jointed by their own collar a few feet below. If all the trusses had been of this design it would have been difficult to categorise the barn separately from the base-crucks. Of the others, three trusses have blades of the full height connected by triangular saddles and an intermediate collar; two have slightly shorter blades with a longer saddle and the blades are imparted a cruck-like profile by the insertion of a curved filler-piece between their inner face and the top collar; and the last, the western most truss, is a cross between these and the upper cruck type, having still smaller "crucks" standing on the upper collar instead of the main one.

There are precedents for variegated trusses in the same building; it is obviously the result of the availability of timbers of particular size and form. Significantly those trusses requiring the shorter blades are the majority. Their distribution however appears to be random. Possibly they were fabricated according to a master design determining only the outline, key levels and profiles, and then erected as and when each was completed. Despite the absence of visible numbers there is no doubt that the trusses were reared from west to east. This is confirmed by the pegs inserted on their west face, though none of them may be original. It is also confirmed by the directional scarfing of the purlins of the lowest tier of the north slope of the roof (there are three tiers on each side). It is only partially confirmed by those on the other side, seven out of eleven being "correct". The upper tiers are likely at first to have been consistent, later replacements accounting for the discrepancies.

That there are no trusses against the gable walls implies no more than that masons and carpenters must have worked more closely together than they did for instance at Middle Littleton. The purlins of bay 1 span from the gable wall across truss I to truss II. Thus truss I must have been erected while the gable wall was being built or very soon afterwards leaving holes for the purlin ends with enough tolerance to push them in and drop them onto their trenches on the backs of the cruck-blades.

The side walls must also have followed very closely the rearing of the crucks, as there is little doubt, to judge by the low level of the oak pads and length of cruck blades bedded in the masonry, that it was the walls which were relied on to keep each truss vertical after rearing until the longitudinal structure of purlins and windbraces had been completed. The porches, too, are fully integrated with the main body of the building. More often one finds the porch roofs are more or less independent of the main roof and could have been added at any time.

*Fig. 19. Bradford-on-Avon barn – north elevation*

Bradford barn, taken into state care in 1919, was very thoroughly restored from 1951 to 1957. The work is typical of the Ministry of Works and now of English Heritage, already seen at Leigh Court. Preservation and permanence of form is invariably the objective, regardless of the means – and cost. Thus massive foundations and concealed steelwork were used at Leigh Court. At Bradford restoration was even more drastic. The entire north wall was rebuilt on new foundations, most of the trusses taken down and repaired with concealed plates and bolts, and re-erected in new cruck pads throughout. Many rafters, windbraces and ashlar pieces were also replaced and the whole structure realigned. The form of the barn is as originally, without exposed or easily visible repairs, but structurally it has been transformed.

## Great Coxwell

This is the largest and finest of the six barns. As an aisled structure it is different from all except Bredon, but at the same time so different from Bredon and all the aisled barns of eastern and southern England that it was clearly the work of a unique carpentry tradition or order of design – namely, as Walter Horn discovered, Cistercian.

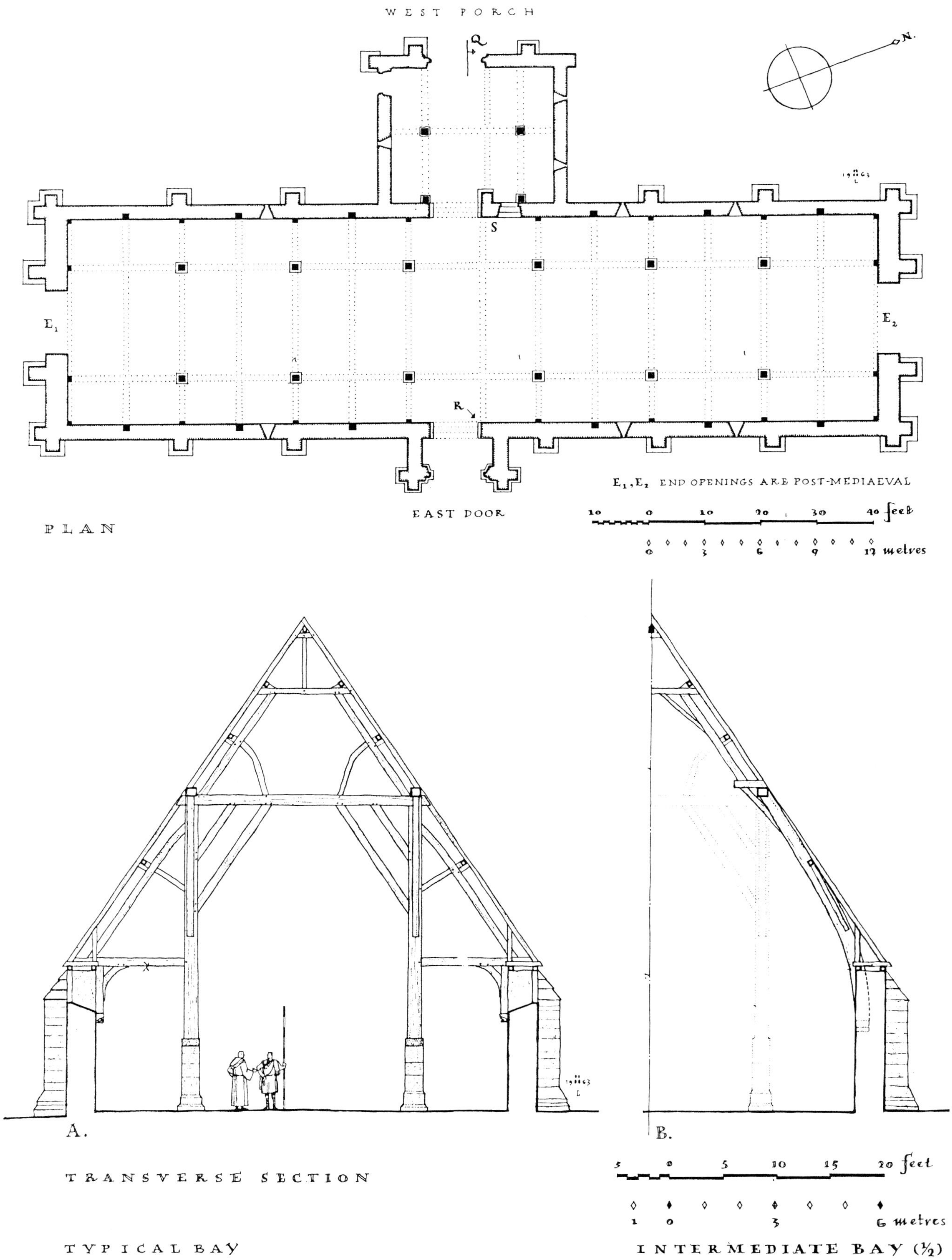


*Fig. 20. Great Coxwell barn – plan and sections (typical and intermediate) by Horn and Born*

*Fig. 21. Great Coxwell barn – interior view, bays 6 and 7*

It is first set apart externally by its single transeptal porch on the west side and lesser porch opposite, instead of the pairs of porches of the other barns. Its west porch also had an upper floor as Bredon but again of quite different design. It was within the envelope of a much larger structure, apparently of one storey, instead of a distinct upper floor built on top of the entry, as at Bredon.

But the chief difference is the height and architectural elegance of its internal framework. Instead of a series of identical bays from end to end, Great Coxwell is designed on the module of a square. In the nave each bay is of the same length as its width, and each aisle bay a double square bisected by an intermediate principal rafter that springs from the side wall and sweeps upward to the arcade-plate at its head. Walter Horn described these principals as base-crucks, but since they are not connected by any kind of tiebeam or collar, they do not form a truss. Each is simply a brace supporting their arcade-plate at its midspan between the posts.

The cross-frames, each standing on a 7 foot high stone plinth, were reared. The posts themselves are 23 feet high so that from floor to arcade-plate it is 30 feet, compared with 26 feet at Bredon. The reason for such high plinths was quite likely to have been that longer posts, if not unobtainable, would have been too top heavy to control during rearing, especially having

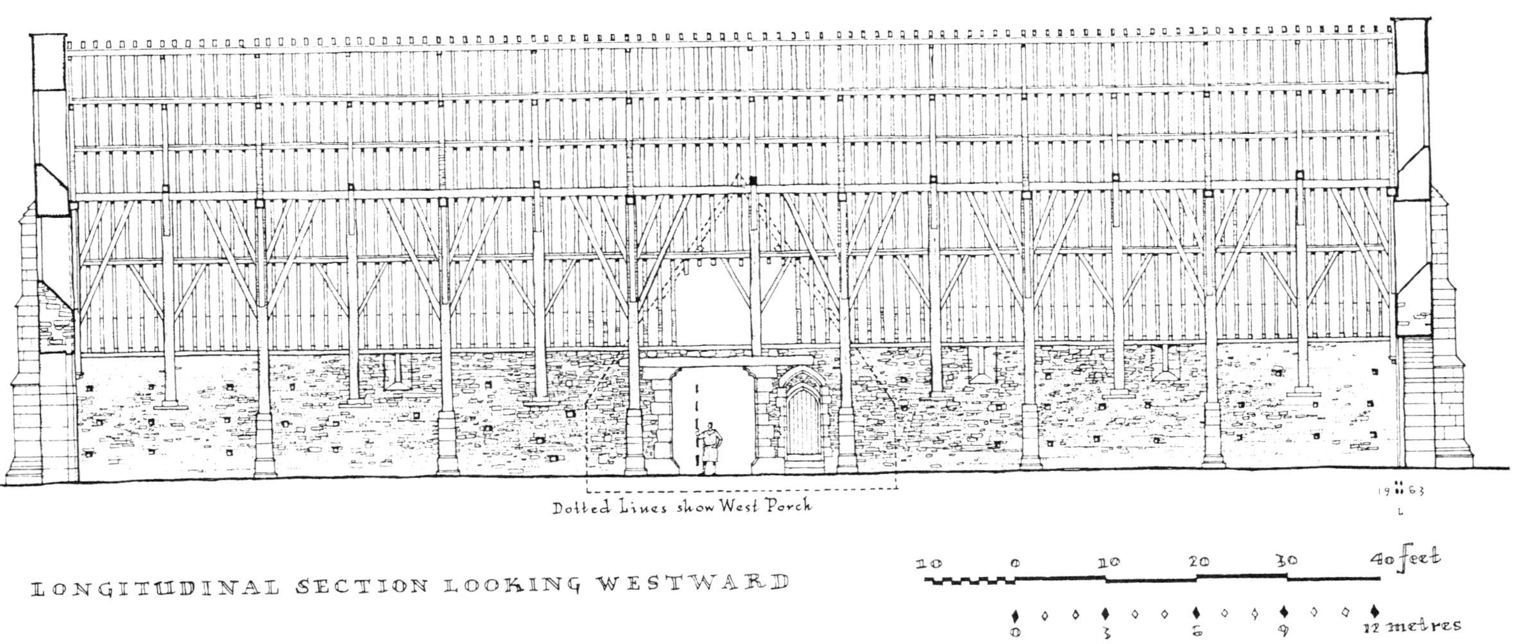


*Fig. 22. Great Coxwell barn – longitudinal section looking westwards, by Horn and Born*

regard to the weight of the double-bracing between posts and tiebeam, necessary to prevent torsion of the post-frame. Rearing an arched frame, such as a cruck, presents no such difficulties. Another feature in which Great Coxwell differs from lowland aisled structures is that it is an example of "reversed assembly". The tiebeam is tenoned to the posts instead of dovetailed over the arcade plate. This of course is necessary to close the frame and make it rearable. The cross-frames of other aisled structures without reversed assembly may still have been reared, depending on whether there was a strainer beam closing the frame somewhat lower down, as at Bredon.

Great Coxwell was restored in 1961 by the National Trust, its owner since 1956. The roof was re-laid, the Cotswold slates re-pegged to sawn laths, correctly not bedded or torched underneath. The Victorian blocking of the great west door - in brickwork of all things – and the wrong reconstruction of the original loft were removed. Undesirable additions to the exterior were also demolished.

Lastly, the most recent dating by dendrochronology by Gavin Simpson of Nottingham University gives 1246. Morris, from the widence of the corbel stones in the side walls wrote that "it was certainly not later than 1250", and Horn comparing the corbels with those at Rievaulx abbey and other Cistercian buildings confirmed it.

*Fig. 23. Great Coxwell barn – corbel stone, by Horn and Born*

*Fig. 24. Great Coxwell barn - view from south-west*

## *Bredon*

The following description closely follows my report to the National Trust of 12th June 1980.

Bredon Barn is an aisled structure of nine bays, with Cotswold stone walls and stone slate roof. The length is 134 feet; its width and height are about equal at 44 feet. There are two wagon porches on the east side, the south-east porch having a loft approached by an external stair. This alone distinguished Bredon from other great barns. It was inhabited by a reeve and his family. It has a fireplace, a stone lavatory basin hollowed out of one of the two window sills, and a garderobe with a four foot square shaft in the re-entrant angle of the barn and north wall of the porch dropping to a pit fifteen feet below.

The stonework is coursed rubble with dressed quoins, buttresses, copings and the loft's chimney, a very rare example with octagonal shaft, side outlets and pyramidal cap. Putlog holes are unfilled and show that the side walls were built with two stages of scaffolding, the north gable wall with nine and the south gable seven.

The side wall buttresses are two stage. The south gable, facing the approach from Bredon village, has a three stage central buttress and two-stage corner buttresses, each making an internal angle with that of the side wall. The north gable, where the ground begins its steep slope down to the river Avon, has a giant four-stage central buttress and two of two stages on either side, but no corner buttresses. The gables have finials with gablets facing all four directions.

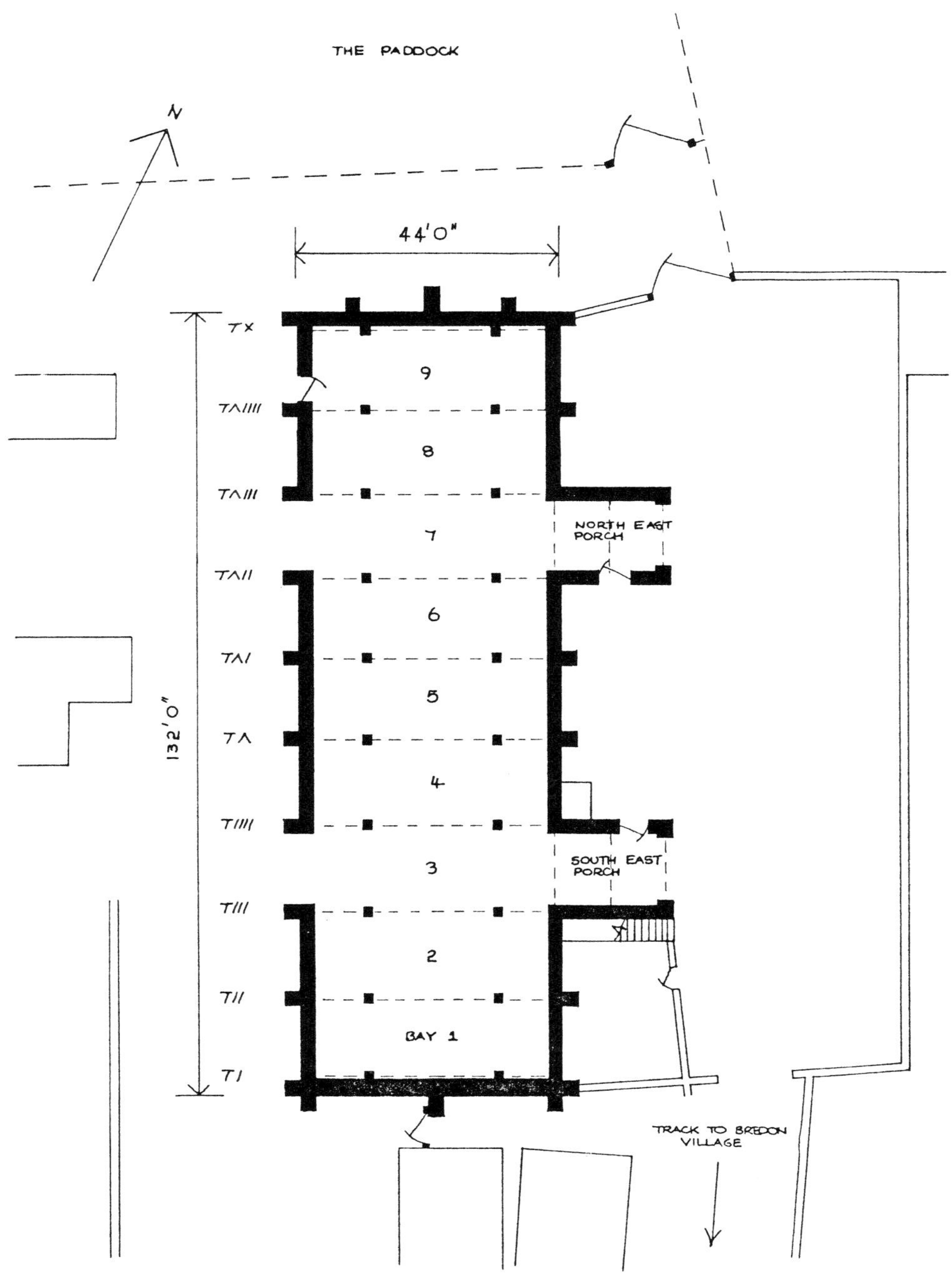


*Fig. 25. Bredon barn – plan showing bay and truss numbers*

The loophole windows of the south gable consist of a lower tier of four (better arranged to avoid the internal timbers than at Middle Littleton) and two upper ones on the centre line, of which the top one is a small cruciform. This gable also has a fire insurance medallion! The north gable has three windows, two lower and a higher central one.

The only alteration to the exterior was the insertion of two doorways in the west wall. Their reinforced concrete lintols probably replaced oak, unless the openings were part of the National Trust's restoration of 1951. The two buttresses of the east wall between the porches have also been rebuilt.

The roof (as all the other Cotswold slate roofs) is laid in diminishing courses with 3'0" slates at the eaves down to 6" at the ridge. The ridge is of 2'6" to 3'6" V-section stones.

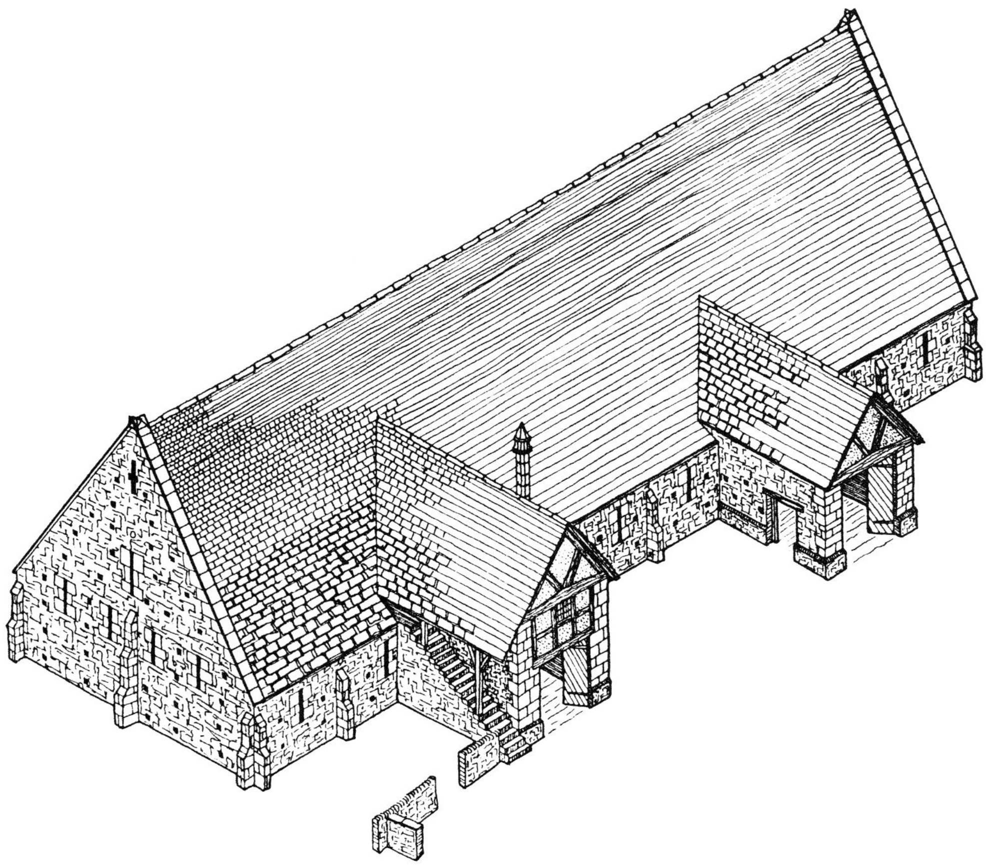

*Fig. 26. Bredon barn – aerial view from south-east*

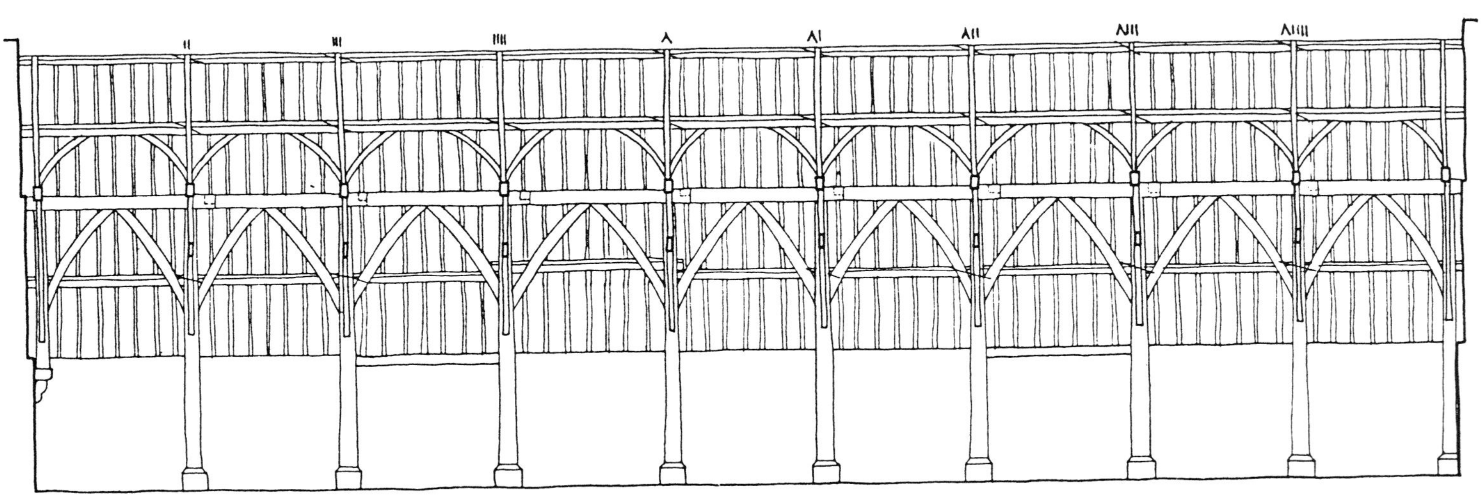

*Fig. 27. Bredon barn – longitudinal section looking west*

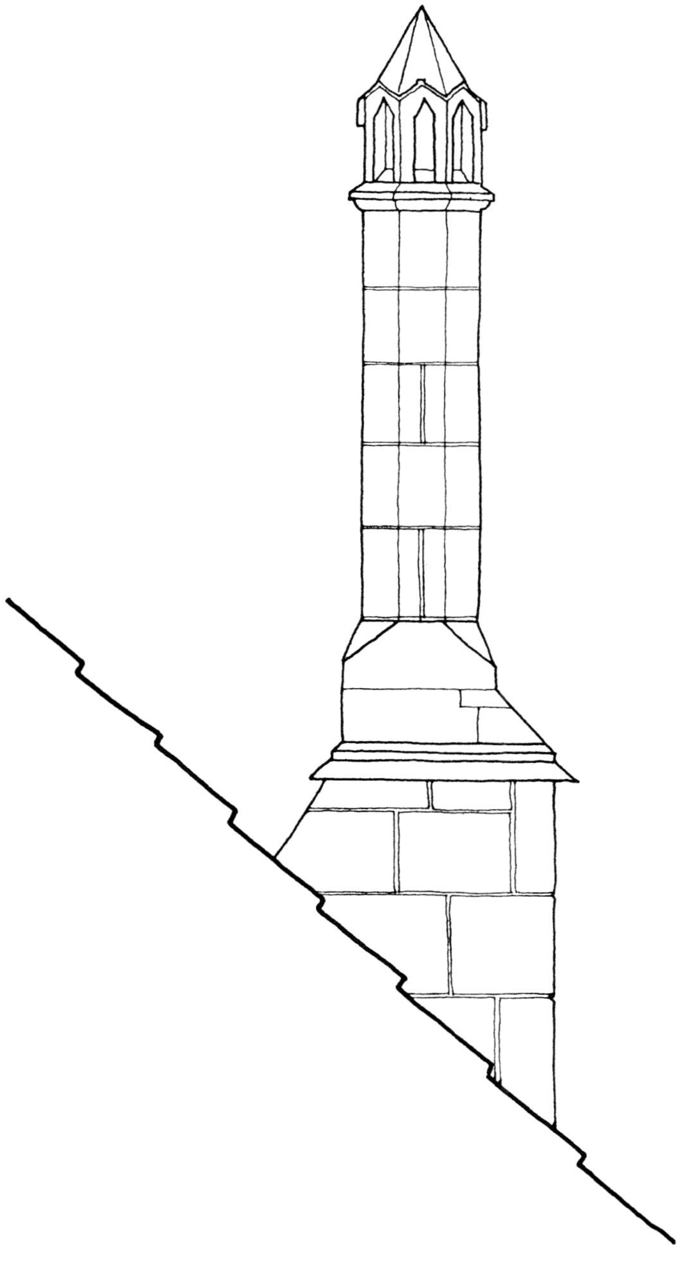

*Fig. 28. Bredon barn - chimney*

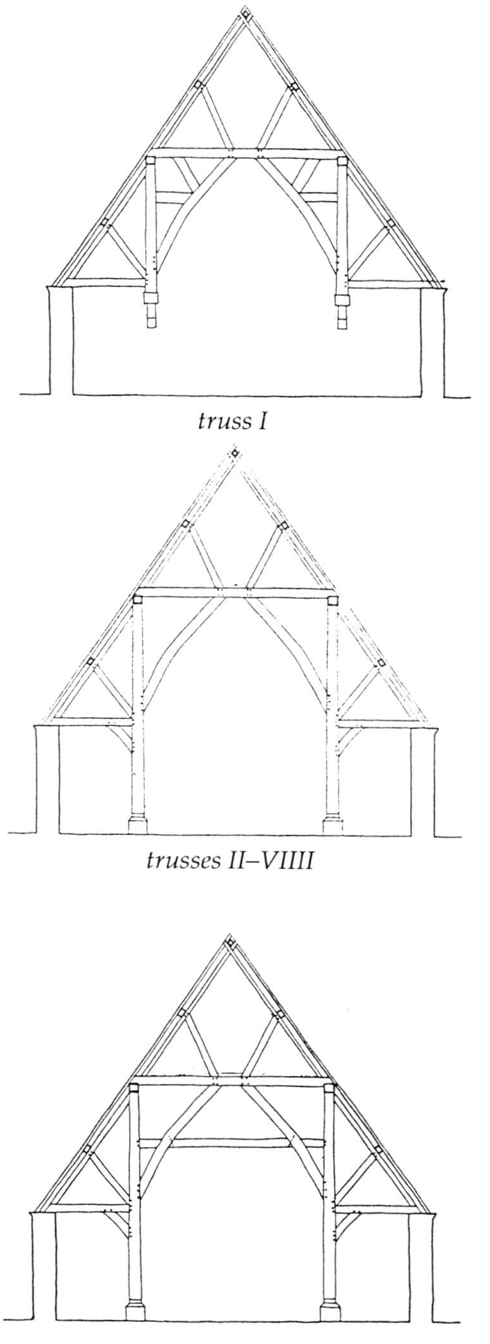


*Fig. 29. Bredon barn trusses*

Internally the timber-framing consists of ten cross-frames with 23'0" high posts on 2'0" square stone plinths, 20'0" apart across the bay and each bay 14'0" long. Their numbering, in $1^{1}/_{2}$" Romans, is complete on the upper face of all the main members of each frame or truss and in correct sequence from I to X. The arcade framing is equally methodical but with numbers at each end of the longitudinal members in each bay. Thus the arcade-plates are numbered I–II in bay 1, III–IIII in bay 2, V–VI in bay 3 and so on to XVII and XVIII in bay 9. The arcade braces are similarly numbered. Truss I was erected piece by piece on the corbels of the south gable at wall-head level, 12'6" above the floor. It consists of two posts, tiebeam and arch-braces with two struts arranged like wheel spokes in the spandrel. Trusses II to VIIII were reared and consist of the two posts, arch-braces and the strainer beam, about 5'0" below the post-head, tenoned into the posts and with halved joints *under* the braces, so that when the frame was reared the braces would also be raised. The tiebeam was not dropped on until the arcade-braces and arcade-plate had been fitted. Truss VIIII, in order to have enough space for framing on the floor, had to be reared *before* the north gable wall had been built. Truss X, on the other hand, designed as truss I, to be assembled vertically against the gable wall, could not have been erected until *after* the wall had been built. The follow-on operations of masons and carpenters were clearly well organised, but there are signs, such as its relative thinness, its absence of offsets and its fewer windows in comparison with the south gable, that this wall was designed to be built as quickly as possible.

*Fig. 30. Bredon barn – interior perspective looking north from truss II*

The aisle trusses, consisting of aisle tiebeams with curved brackets to the posts, compound principals and inclined strut from post to principal were assembled after the rearing of each truss and as the side walls reached wall-head level. The aisle purlins were placed and held by a short strut between the two rafters making up the complete compound principal.

The upper roof of the nave has the same design of principals; the lower member intersects its opposite number at the ridge to seat the ridge pole of diamond cross-section.

There are windbraces only in the lower tier of the nave roof. There are four tiers altogether all of equal height, defined by the purlins; the rafters of the aisles are the same length as those of the nave. Thus, in plan the width of the nave is twice that of each aisle, less the wall thickness. The porch roofs have the same design of compound rafters, but in the open roof of the porch loft the intermediate truss is arch-braced, without a tiebeam.

The only structural alteration to the framing was replacement of some of the purlins in the west aisle. The replacements were lapped instead of bridle-scarfed giving an up-and-down appearance, out of keeping with the precise jointing of the other horizontal members, particularly the arcade-plates. Other alterations were rafter replacements some of which were sawn softwood, no less out of keeping than the purlins.

Otherwise the completeness of original members and their framing was remarkable, no doubt largely due to the quality of the timber throughout, the precision of its construction and the stability in the design of the whole building. In this respect it outclasses even Great Coxwell, but architecturally Great Coxwell is the more elegant with its soaring roof and wider spaces. The construction of all the other barns is so different that comparison ends with size.

Lastly, the dendrochronological samples taken in 1969 failed to date the barn, but further samples from a much wider range of timbers were selected by Dr Fletcher of Oxford in 1982 during reconstruction, and this yielded firm dates for felling the trees from 1335 to 1345.

*Fig. 31. Bredon barn – south-east porch from north-east*

*Fig. 32. Bredon barn – view from south-east*

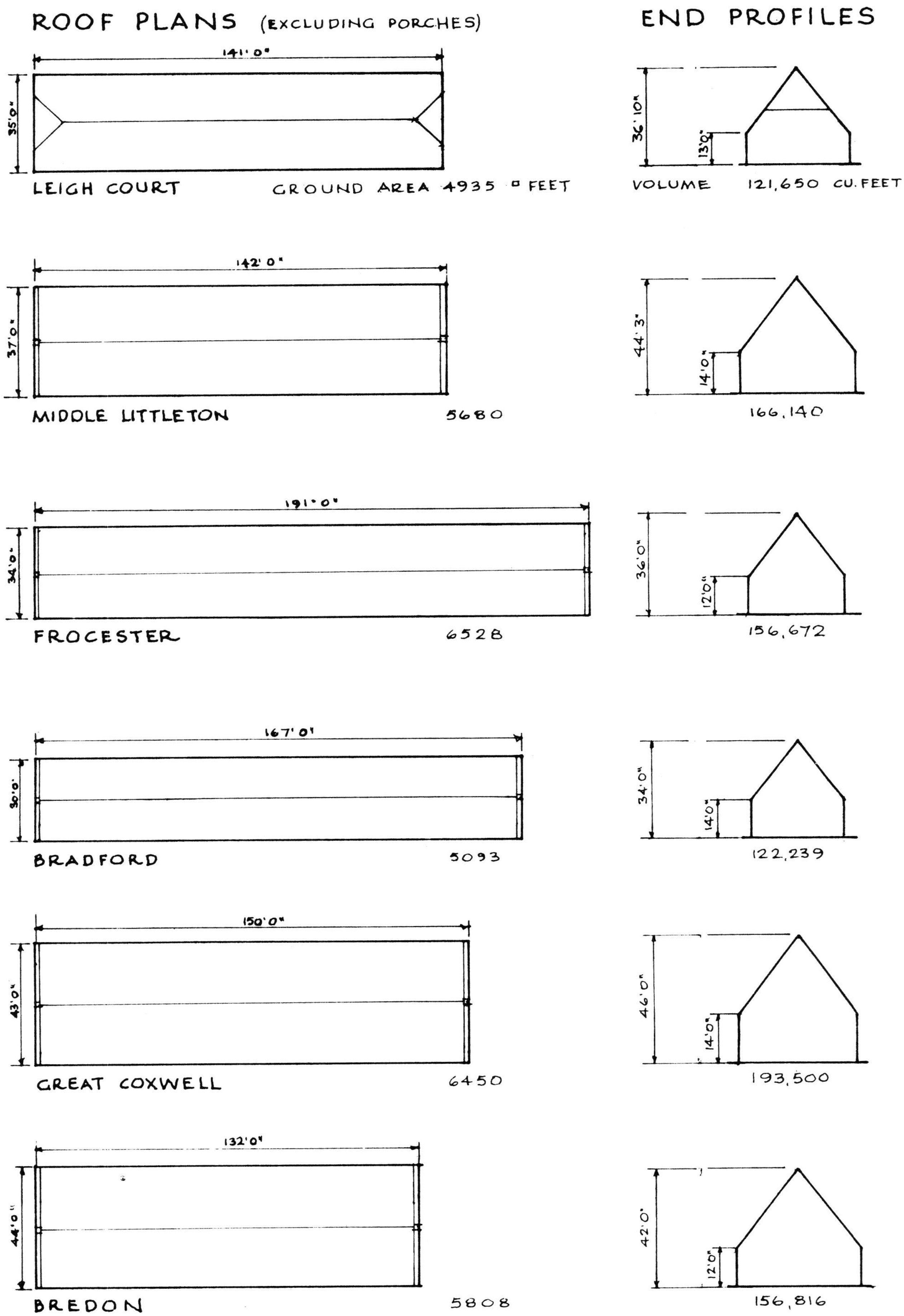


*Fig. 33. Comparative dimensions of the barns – plans and end profiles*

# Chapter 2

# BREDON BARN IN HISTORY

*by Christopher Dyer*

The construction, use and survival of a great medieval barn are important historical events, and historians ought to be able to place such a building in the context of its place and time. In the case of the Bredon barn, we are helped by the survival of a quantity of documents from the estate records of its builders, the bishops of Worcester. Here we need to answer the obvious questions: why was the barn built? And how was it used? I will also try to explain why it has survived.

In answering the first two questions, we must consider the controversial problem of 'tithe barns'. Bredon, in common with many large old barns, has frequently been described as a tithe barn, and indeed that phrase is still used by the National Trust. In order to explain why this is an error, it is necessary to sketch the origins of farming and lordship in Bredon. Communities who used the local resources, and especially the good arable land in the river valley, lived around Bredon throughout the Iron Age, the Roman period, and the early middle ages, and this rich land attracted the wealthy and powerful who wished to draw incomes from the agricultural activities.[1] In the eighth century we find Bredon becoming the site of a prominent monastery under the patronage of members of the royal family of the kingdom of Mercia.[2] The normal convention is to call it a 'minster' because it was a rich church staffed by a group of clergy, not a community of cloistered monks living under a strict rule. The clergy of the minster would have served as priests to the local population, but the minster also bolstered the prestige of powerful patrons, who made it very rich with grants of land not just in the immediate vicinity but as far away as Cutsdean in the Cotswolds, and near Alvechurch in the northern woodlands of Worcestershire. In the ninth century, as commonly occurred, it came under the domination of the bishops of Worcester, and the lands of the minster were ultimately divided between the local clergy, whose successors were the rectors of Bredon parish; and the bishop, who took the lion's share of the endowment of the minster as part of his huge landed estate.[3] The rectors were very rich – at the end of the middle ages they enjoyed an income of about £70 per annum, when the average parish clergyman had only about £10 annually.[4] These revenues came from a large glebe of about 200 acres, but also from the tithes of agricultural produce from the very large area which had originally belonged to the minster – not just from Bredon and its neighbouring hamlets of Hardwick, Norton and Westmancote, but also from such outliers as Cutsdean. The tithe corn of a cultivated area of thousands of acres cannot have been much less than 150 quarters of grain each year, and would have needed a substantial storage space. However, the rector's tithe barns were attached to the rectory itself. Two large barns are described in the glebe terrier of 1616, and a large building is shown on the map of 1811, adjacent to the rectory, to the north of the church.[5]

The surviving Bredon barn which is the subject of this book stood between the manor house and the manor farm as is clearly shown on the map of 1811.[6] It was a manorial barn to store the crops produced on the bishop's demesne.

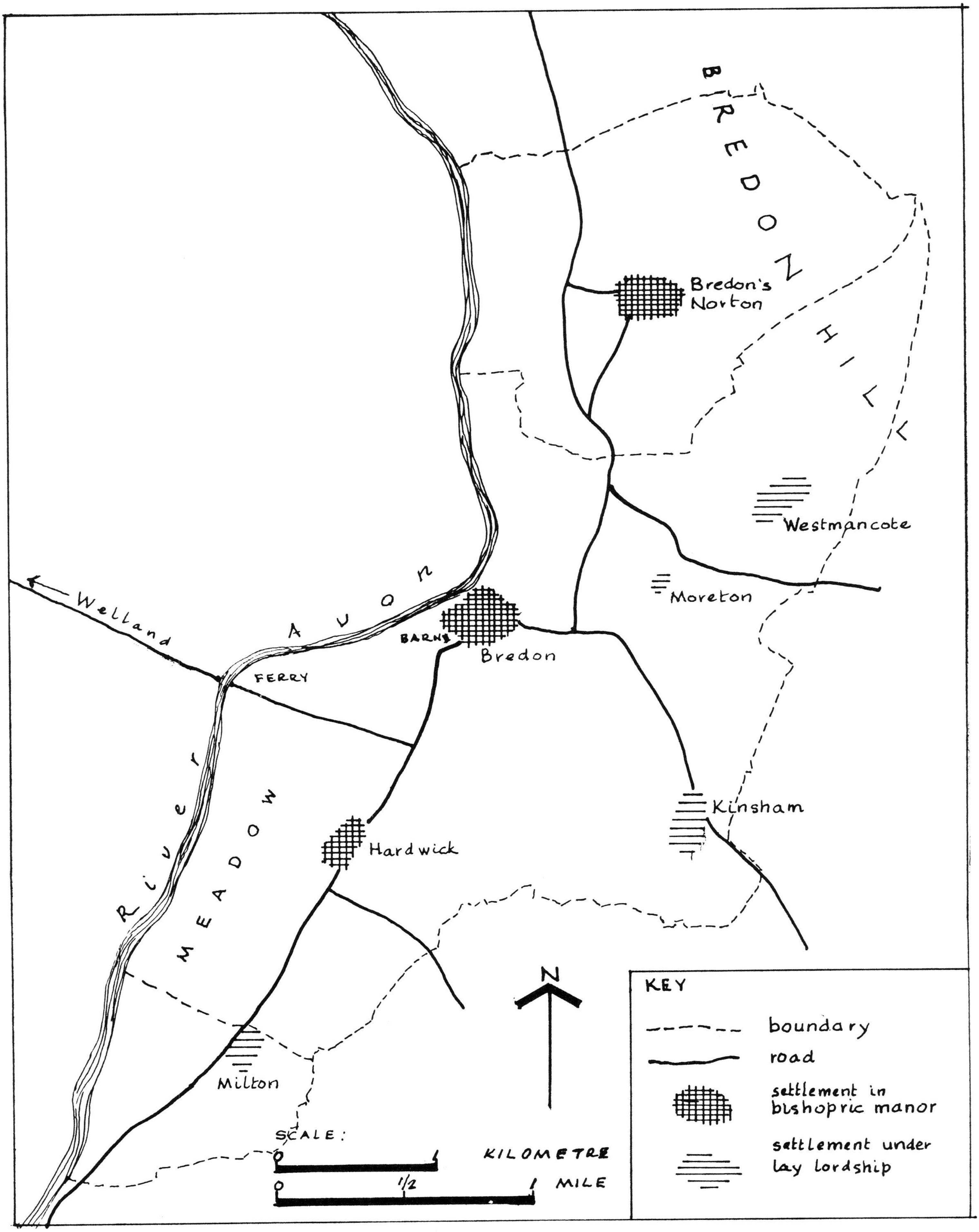


Fig. 34. Bredon manor

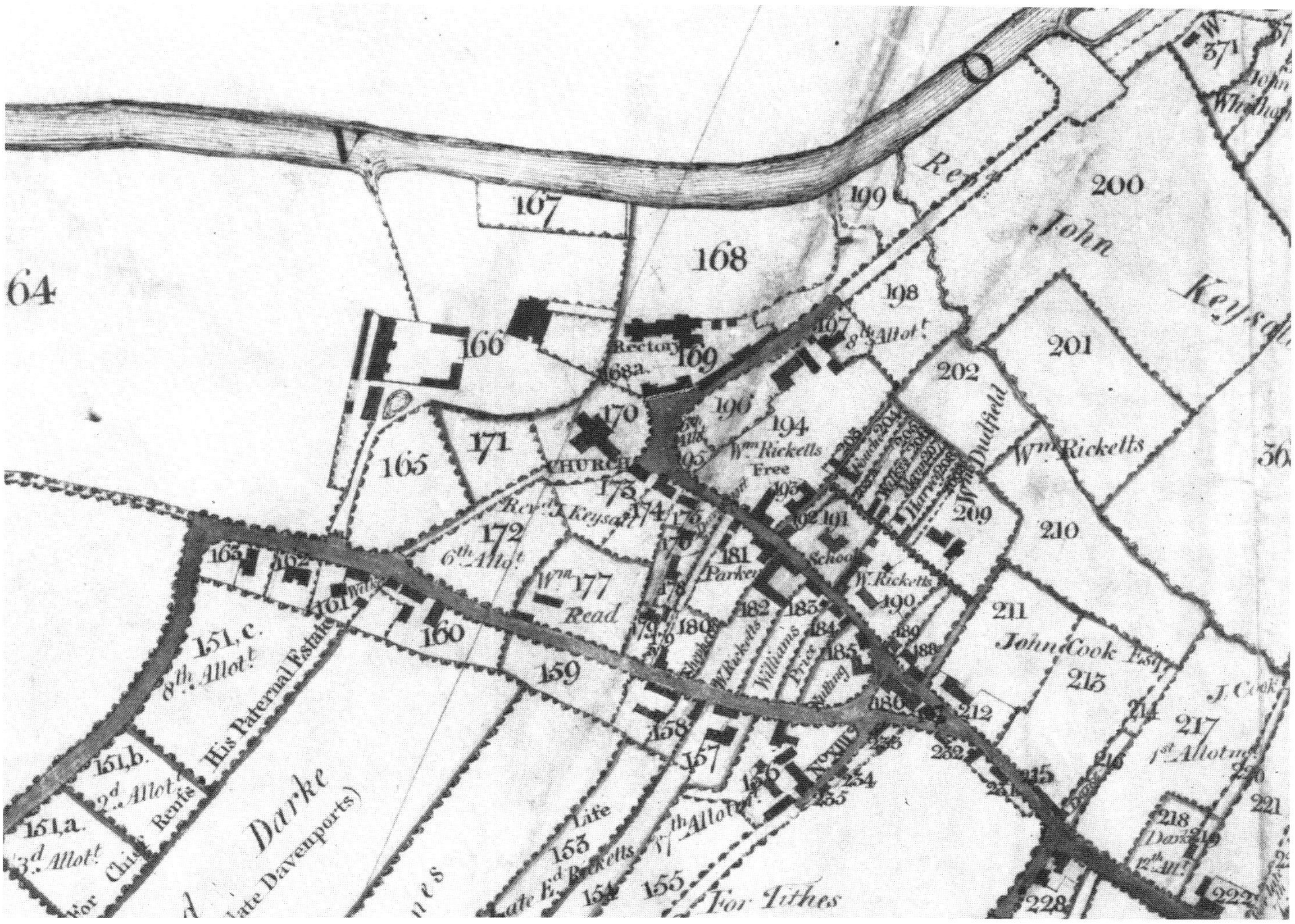


*Fig. 35. Part of enclosure map of 1811. Bredon barn is extreme top-left (166), the Manor House is right (last), and the rectory to right again, with its large barn opposite (169)*

In short, the lordship of Bredon and its dependencies which had been in the hands of the minster in the eighth century was divided in the ninth between two ecclesiastical organisations, the bishop and the rector, and these two institutions survived until the reformation with a quite separate administration, and a separate set of buildings; our barn belonged to the bishop and was not a tithe barn. It should be said that this strict distinction between the bishop's assets and those of the rector were occasionally blurred. The bishop could have become the recipient of the tithe income from the parish of Bredon if he had appropriated the rectory. However, this was a public and formal act which required elaborate documentation (as in the case of the appropriation of Blockley rectory in 1335 by bishop Montacute), and this did not happen in the case of Bredon.[7] In 1352 bishop Thoresby leased the rectory of Bredon from the rector for five years, presumably in order to increase his revenues after the Black Death.[8] In that short period the bishop did collect the tithes, but it was only a temporary arrangement and in any case the bishop is likely to have used the existing rectory barns, which would have come with the lease, to store the tithes. In the sixteenth and seventeenth centuries, we find the lord of the manor sometimes farming the rector's tithes, so again it is conceivable that tithe corn was then stored in the barn, but this would have been only an occasional use.[9] The barn was built to serve the manor. This point needs emphasis because a common fallacy is to describe any old, large barn as a tithe barn, and we should be more precise in our language, because a very high proportion of surviving medieval barns were in fact of the Bredon type and formed part of the complex of manorial buildings.

Once we have established that our barn was attached to the bishop's manor of Bredon, we can begin to explain why it was built. The dating of the barn's timber from the evidence of dendrochronology places it in the middle years of the fourteenth century, probably the 1340s.[10]

The organisation and economy of the Bredon manor are well documented in three surveys produced between about 1250 and 1299.[11] These show that the bishop had in Bredon a manor house and its associated buildings; a demesne of just under 400 acres of arable land lying in the open fields around the village of Bredon; 100 acres of meadow in the low lying land next to the river; rights of pasture on Bredon Hill, including summer pasture for sheep; and access to the bishop's wood at Welland some seven miles west of Bredon on the lower slopes of the Malvern hills. Labour to work the demesne came from the hundred or so tenants who lived in the villages of Bredon, Norton and Hardwick. Many of these were customary tenants owing labour services, though it is likely that by about 1300 that a high proportion of these services were commuted into cash payments, and most of the labour on the demesne would have been carried out by wage workers recruited from among the cottagers and smallholders of the three villages. We can recognise, from the surveys of the late thirteenth century, how an ecological balance was achieved in the agricultural management of the bishops, because in addition to extensive arable cultivation, they used the hay and lowland grazing to feed a herd of cattle, and were able to provide winter quarters for sheep flocks from other manors, and also kept a substantial number of pigs, which were fed on some of the produce of the demesnes, particularly peas, but were also driven to the woods at Welland when the acorns were available in the autumn. However, our focus must be on the production of cereal crops because it was for that purpose that the barn was built.

Why was the decision made to build the barn, probably in the 1340s? There would of course have been a barn built in the thirteenth century, perhaps even in the twelfth, which was thought to be in need of replacement. The decade of the 1340s does not mark a very high point in the history of the estate, nor indeed in the agricultural economy of England generally. The bishops of Worcester had done very well in the thirteenth century. They had tripled their income from their estates from £400 to £1200, by profiting from their demesnes through sales of produce, and by squeezing their tenants to pay more rent in cash. The peak of this expansion had been reached by c. 1310 and in the early fourteenth century the estate was having some difficulties.[12] There were complaints that infertile lands and poverty among the tenants made it difficult to maintain estate income at a high level. One response to this problem was to reduce the size of the demesnes and to cut the costs of production, by renting out land for a fixed rent. Indeed this happened at Bredon in 1344 when 78 acres of the demesne arable were leased to tenants.[13]

There is probably no paradox that the barn should have been built at this time. We could see the barn as an investment on the part of estate managers who wished to improve the efficiency of their farming, by preventing waste and ensuring that crops were kept dry and secure. In making the decision to build the barn they may even have been anticipating that in the long term they might need to lease out the whole manor, and that they would gain a better return from a farmer or lessee if the demesne was well equipped with good quality buildings. Bredon was certainly not alone among manors in the early fourteenth century in its expenditure on new buildings.[14]

We must envisage the decision to build the barn being taken by a group of managers, gathered at the bishop's palace at Hartlebury or Worcester, perhaps even under the chairmanship of the bishop himself. It is likely that the builder of the barn was Wolstan de Bransford who was bishop in 1339–49. His advisers would have included Peter de Grete, the steward of the estate (its chief official), who was himself the lord of a nearby manor at Pirton and well aware of the economics of agriculture, and would be generally well informed about financial matters from holding office as sheriff of Worcestershire.[15] No doubt the receiver or treasurer of the estate would also have been present, knowing as he did the precise resources available for expenditure on the barn. A likely candidate would be John le Botoner, appointed as receiver in 1342, who was also rector of Naunton in Gloucestershire.[16] Once the decision had been made, the construction work would have been supervised by local officials, the bailiff who would have looked after a group of manors including Bredon, and the reeve who was primarily responsible for the running of Bredon manor itself. Whereas the bailiff would have been an administrator of some status, often coming from a minor gentry family, the reeve would have been a customary or servile tenant temporarily put in charge of the demesne.

The barn was built to high specifications. It had a capacity for 420 quarters of grain in sheaf, which would have been rather more than was necessary for the produce of the demesne, and it was probably intended to house hay as well.[17] It was built by skilled workers – masons, carpenters, slaters etc. At this period they would have received about 3d per day, and hundreds of man days would have been needed, probably spread over a period of two years. In the absence of detailed records, one must guess from contemporary parallels that it required expenditure near to £50, which was the bishop's annual revenue from the manor of Bredon for one year.[18] In two years the bishop may therefore have invested close to a half of his profits on this project. Later documents from the 1370s onwards which record building activity at Bredon give us the likely sources of the materials. The building stone would have come from quarries on Bredon Hill itself. The major timbers were not available from adjacent woods, but would have been transported from Welland across the Severn. The Cotswold slates came from specialist quarries 12 miles away at such places as Snowshill and Temple Guiting. The lime for mortar, nails and other specialist materials would have been bought at Tewkesbury, the nearest town.[19]

The documents of the late fourteenth century show us how the barn fitted into the workings of the manor. When it was built it formed part of a great complex of structures which fell into two groups. On the eastern side of the manorial establishment stood the bishop's manor house, a major residence with a hall for meals and public occasions, at least six chambers for the bishop and his servants, a chapel, and stables for their horses. In order to serve food and drink for the bishop's household which consisted of at least 20 or 30 people there was a kitchen, brewhouse, bakehouse and 'dresser' for storing meat. On the western side of the manor house lay the barton in which stood the barn, a byre for oxen, a cow house, stable, sheepcote, pig stye, poultry house and dovecote.

The bishops would only have visited Bredon occasionally because they led an itinerant life. For example, in 1344–5, Wolstan de Bransford, the likely builder of the barn, spent the winter at his principal residence in Hartlebury, then moved in the summer between May and July to Bredon, and travelled on in August to Henbury-in-the-Saltmarsh near Bristol, returning in the autumn to Withington and Blockley in the Cotswolds, and then back to Worcester and Hartlebury.[20] If a bishop visited and stayed for as long a period as Bransford did that year, then the principal function of the barn would have been to store crops for the sustenance of the household – wheat for bread, barley malt for ale, and oats for the horses. We can see the barn as playing an important part in a regime of self sufficiency in which the bulk of the crops from the estate would have been consumed directly by the lord, his servants and their horses.

But in some years the bishop did not come, either because his itinerary took him elsewhere in the diocese, or because some bishops, unlike Bransford, were absentees who had important business elsewhere in the kingdom or abroad. In such circumstances, the produce would have been sold, and the barn then become part of a system of commercial grain production, in which the various crops were threshed, and the grain was taken to Tewkesbury. There it could be sold to a cornmonger, who would then no doubt transport the grain by boat down the Severn to Bristol or even further. At least some of the grain stored in the Bredon barn in its early years may well have found its way to Bordeaux where specialisation in wine production created a demand for imported grain. In such circumstances the optimum time for selling the grain would have been in the early summer, well after the previous year's harvest, when the price would have been most advantageous. The fact that the bishops and lords like them had such ample storage facilities gave them an advantage over peasant producers in being able to keep the grain for a long time and therefore to obtain the best price. It must also be remembered that whether the surplus of grain from the manor was consumed by the household, or sold, that a high proportion of the grain from the manor was consumed internally. Farm servants and visiting workers were paid partly in grain, and a good deal of the peas and oats in particular were fed to the animals on the manor.

The different roles of the demesne in which the barn was involved can be illustrated by comparing the crops at Bredon and their use in 1384–5 and 1395–6.[21] In the first year the manor gained from the demesne (with a little from peasants, because they paid churchscot in the form of wheat), 44 quarters of wheat, 8 quarters of rye, 75 quarters of barley and dredge (a mixture of barley and oats), 36 quarter of peas and beans, and 25 quarters of oats, a total of 188 quarters (a quarter weighed about 500 lbs). 41 quarters of this went as pay to servants, officials, and visiting workmen, all of whom received some of their wages in kind; a little more than this went back into the ground as seed corn; 9 quarters were fed to animals; and of the remainder, 1 quarter 3 bushels was sold, and no less than 87 quarters went to provision the household. This was in the time of a bishop who resided in the diocese, Henry Wakefield. In 1395–6, when the bishop was Tideman of Winchcombe, a courtier of Richard II and therefore often absent from Worcester, the total income in grain was 132 quarters. Again almost a half went on seed and on paying employees. 12 quarters were used to feed animals, and of the rest, the household was supplied with only 13 quarters (mostly wheat), and 44 quarters were sold. The carriage of the grain to Tewkesbury cost 4s., and it fetched £6, presumably from Tewkesbury cornmongers who we know from other sources were operating on a large scale in the late fourteenth century.[22]

Insights into the use of the barn are provided both by the manorial accounts, and by the survival in a remarkable state of preservation of the chamber above the south door, which links up with documentary evidence for the work of the reeve. We must envisage the reeve using the chamber as an office, where he could supervise the estate workers in the harvest season bringing cartloads and wainloads of corn in sheaf into the barn and stacking it. Throughout the rest of the year he would be able to observe the process of threshing – a long and laborious task – and the transfer of the threshed and winnowed corn to a nearby granary. By the late fourteenth century the office of reeve had become a long-term possession of one family. John Frygg served as reeve in 1375–85, and was succeeded by Richard Frygg (who may have been his son) from 1385 to 1399. They were prosperous customary tenants who clearly had learnt the management skills which enabled them to run the manor to the lord's satisfaction, and no doubt with some profit to themselves. In the cupboard of the chamber they would have collected the tally sticks and slips of parchment, in order to provide them with the evidence of their various transactions during the year (rent collection, sales of produce, expenditure on the running of the demesne). These documents would have furnished the information from which the annual manorial account would have been written, later to be checked by auditors.[23]

Specific references to the barn in those accounts show that it was subject to the usual running repairs. It needed two locks with keys for two doors at a cost of 14d. In 1384–5, and in 1392–3 a carpenter was hired to make a 'foil' for the great door of the barn. In 1393–4 four boards were used to build a cupboard in the reeve's chamber, and a mason repaired the fireplace also in the reeve's chamber, which must refer to the existing room above the south porch. There is also a reference in that year to temporary structures within the barn – *floryng* was made from broken carts and wains – a rather rough and ready temporary structure, for what purpose is not revealed. There are constant references to the carrying of the lord's corn and stacking it in the barn, and hay was dried in the barton preparatory to being taken into the barn.

Why has the Bredon barn survived when so many contemporary buildings are lost? One likely reason for its survival is the high quality of the initial construction and materials. But there is no reason to believe that it was exceptionally well built in comparison with barns on other manors in the vicinity which have now gone, and we must take into account as a factor in its preservation the later management of the demesne of Bredon in which the barn continued to play an important part. In 1401 the Bredon demesne was leased to farmers, a normal procedure at that period when lords found the profits of arable farming were reduced as the cost of wages rose and grain prices diminished.[24] The lease included the barton as the Bredon demesne was leased as a single unit, which meant that the main buildings were retained and used by the farmer. At many other manors the demesne was broken up into smaller parcels, which might make the original central buildings redundant. The farmers who took over the management of the demesne had problems in making profits, just as the bishop's managers had done in the late fourteenth century, and buildings, like the barn, were in danger of neglect and decay. However, in 1457 the bishop allowed the farmer a remission of rent totally 24s 7d to repair the stone wall at the end of the barn – presumably the northern gable – and this quite expensive and major work may have helped to preserve the barn through a difficult period.[25] In modern times the manor farm was divided, but only into two parts, and evidently one of the farmers could continue to put the barn to good use, and so it has survived as a major asset for the agricultural management of Bredon into recent times.

The Bredon barn is historically important because it tells us a great deal about medieval society, economy, and technology. The size and solidity of the barn points to the great social inequalities of the period. One must remember that there were perhaps 100 peasant barns in the villages of Bredon manor in the later medieval period of which none have survived. The technical achievement represented by the barn was not a product of the intellectuals of its period, but of the practical skills of its builders, basing their methods on accumulated wisdom and trial and error. The bishops presumably gained considerable benefits from the quality of its building in terms of the safe storage of crops. The assembly of the materials and labour for the building of the barn tells us a good deal about medieval people's ability to make use of resources, not just from a narrow locality but over a considerable distance. The transport system, contrary to popular mythology, was capable of moving heavy loads like great timbers over many miles of road. The use of the barn again is an indication of the ingenious way in which the arable, pastoral and woodland ecology of the region were brought together to maximise the profit from the manor. The decision to invest in the barn, and its subsequent use by the reeve, working on his accounts in his chamber, provides us with some insights into the complexity of medieval administration and the way in which assets were managed. The survival of the barn symbolises the continuity of rural life. Here we see a building constructed at the end of the boom of agriculture in which production was managed directly by landlords, but which continued to serve the need of the demesne for centuries after it was put into the hands of farmers.

## Chapter 3

# THE FIRE

Sub-officer Christopher Parsons of Pershore Fire Brigade, who was in control of operations at the fire, has written a report for this chapter. He says:

"Bredon Barn was used as a working barn and as such there was always the risk of fire. Local residents were concerned about the way the barn was being used and this concern was communicated to the Fire Brigade. The Fire Brigade instructed a crew from Pershore to inspect the barn and its contents. This was carried out at the beginning of April 1980. The barn was found to be full to capacity with hay and straw from previous harvests. It was also used for the storage of farm vehicles. A rick tester ( a long pointed probe with a thermometer attached) was inserted deep into the contents of the barn and a temperature which gave the Fire Brigade some concern was recorded. The results of this inspection were passed to the owner and user of the barn. While in the vicinity of the barn the Fire Brigade took the opportunity to check the availability of water supplies in the area around the barn.

*Fig. 36. Bredon barn – looking down the barn from post IIII; the fire extinguisher is on post VIII*

In the afternoon of 18th April, 1980, fire broke out in the barn. The fire is believed to have been started by one of three boys, returning home from school in Tewkesbury, playing with his father's cigarette lighter. A visitor to the farm heard the boys' cries (luckily they were safely out of the barn). He went to investigate and upon seeing the fire he was of the opinion that if only he could have found a bucket of water, let alone a fire extinguisher, he could have put the fire out. There was, in fact, an extinguisher attached to post VIII, next to the north-west doors where the boys had entered the barn, but it was hidden by hay and straw.

The visitor had to find a telephone in Bredon village, an area which he did not know. This took valuable time, time in which the fire developed rapidly.

At 16.22 Fire Brigade Control in Worcester received a call reporting a fire at Bredon Barn. The call was passed to Worcester by the Gloucester Fire Brigade Control. This was because all telephone lines in Bredon go through the Gloucester exchange even though the village lies in Worcestershire. Fire appliances from Pershore and Evesham were mobilised to the incident immediately. Gloucestershire Fire Service also sent one appliance from Tewkesbury. Each of these appliances was carrying in excess of 400 gallons of water which can be brought into use in a matter of seconds. The distance from Pershore to Bredon is eight miles and after travelling only two miles a plume of thick smoke was visible on the horizon.

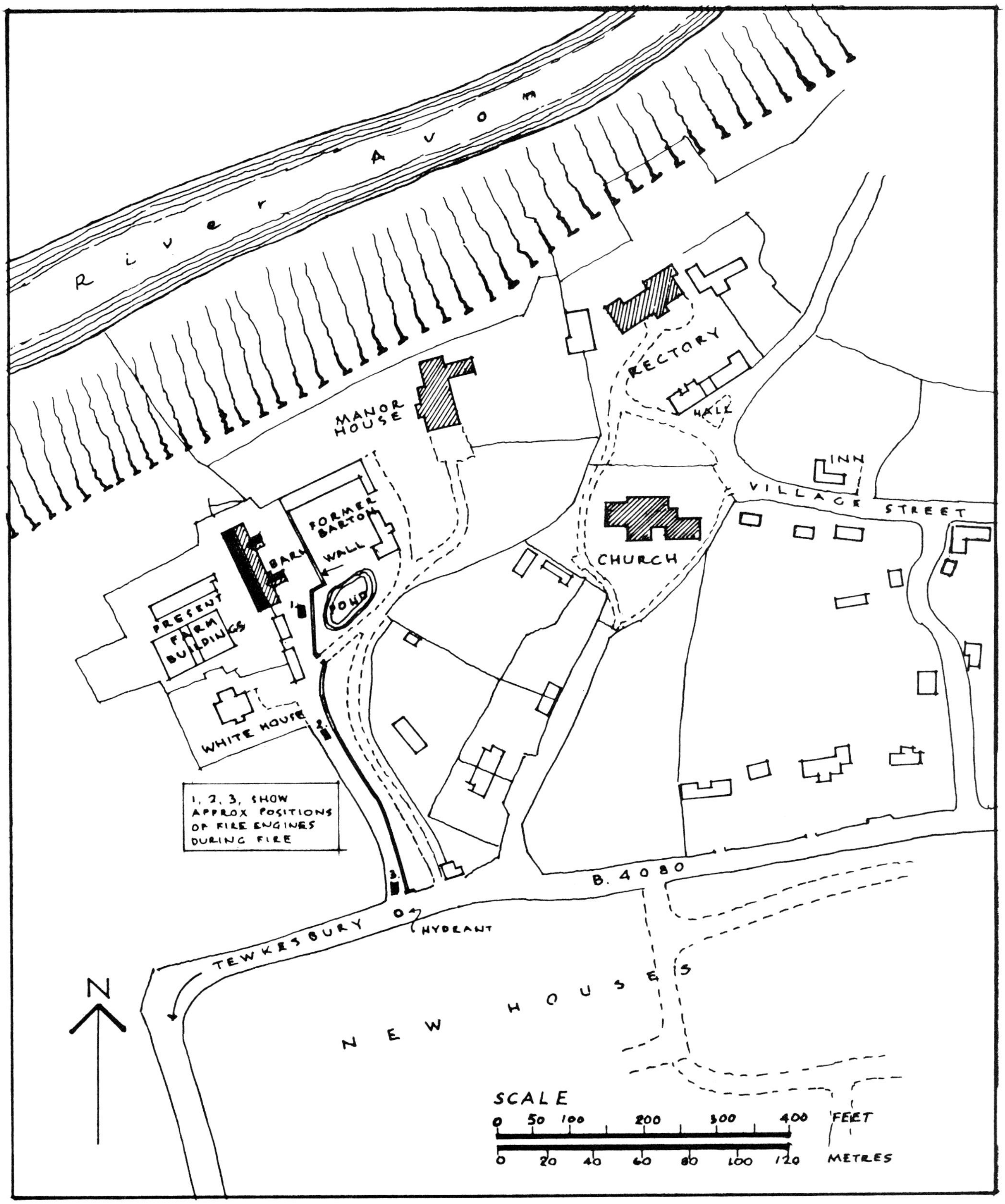


*Fig. 37. Map of Bredon showing barn and important buildings and positions of fire engines during fire*

The first fire appliance arrived at the scene at 16.30 (only eight minutes after receiving the call) followed by a second appliance some four minutes later at 16.34. The building and contents were found to be well alight with flames already through the roof. The situation was quickly assessed and a decision made to concentrate all immediate efforts to save as much of the building as possible.

Due to the ferocity of the fire and the limited working area immediately around the barn the fire appliances were parked in the lane leading up to the barn. To park fire appliances out of danger and provide a good working area is normal fire-fighting practice. Jets, using water from the fire appliances, were played onto the structural timbers of the barn together with cooling jets on the stone walls.

I immediately sent a radio message to Fire Brigade Control in Worcester giving details of the building, containing approximately 200 tons of baled hay and three farm vehicles, being well alight. The radio message also detailed the number of jets that were in use, where the water was being taken from and also asked for a further fire appliance to attend. This was required to draw water from the river Avon and pump it up a steep bank to the fireground. With so many jets being used the water in the fire appliance would only last a few minutes. Additional supplies of water were taken from the water main in the road and also from a large pond in the garden of the Manor House. Firemen were hampered by the high wall between the fireground and the pond. Despite these difficulties at no time did the firefighters run out of water.

When the fourth fire appliance arrived the crew carried portable pumps down the steep bank to the riverside on the north of the building. Hoses were laid out to the fireground and a plentiful supply of water was provided. This enabled the hydrant in the road to be removed as it had been seriously hampering the passage of traffic. Firemen who were not handling the jets were detailed to remove the burning hay from the barn. To do this by hand was a long and exhausting task. The objective was to reach the upright timbers of the barn as quickly as possible and remove the hay from them. This would assist in maintaining the integrity of the building.

The fire-fighting operation continued as described above and at 18.08 a message was sent to Fire Brigade Control stating that no further assistance would be required at the incident.

For the twenty-four hours after the fire, the Fire Brigade carried out further inspections of the building to ensure that no small pockets of fire had been overlooked."

His report does not mention the general circumstances of Bredon Barn either as leading up to the fire or in fighting it. First, the trust had acquired the barn in isolation from its Manor House and traditional surroundings in 1951. The stone wall was built leading from the main road to the barn, and the lane was for visitors only. Also the roof slates were relaid.

The barn was rented for thirty years to a farmer who lived on his other farm about ten miles away, visiting Bredon only when there was work to be done. His access to the barn was on its west side, where anyone could get in unseen except when he was there. The lease had one more year to run. Whether he left the doors open or not, they were openable, and only the private house standing between the barn and the road and whose owners had no connection with the barn had to be circumvented. This was hardly difficult as a footpath led off the lane just before reaching the barn. Thus, the question is not so much: why was Bredon burnt, as how had it survived so long? Only fire engines could not reach the west side.

Nor could the firemen even get into the north-east porch as its two double doors and the side door were blocked on the inside by the three farm machines noted by Parsons. Only the south-east porch could be entered, and, very significantly, this and the two adjacent bays, 3 and 4, were saved. Had they gone there would have been no hope of reconstruction. A further question is raised: if the north-east porch had been open might not it and its adjacent bays, 6 and 7, also have been saved? And finally, as the Worcester Fire Officer explained to me a few days later, the flames which started at the north end caused such a build-up of heat in the top roof that they swept through its whole length utterly destroying it and damaging the stonework of the south gable more severely than that of the north. The slates, he noted, had been bedded in mortar. Had they been laid dry, a more difficult and skilful method, the heat and flames could have got out and the fire might have been confined to the northernmost bays, 8 and 9.

Also as a result of bedding, the slates fell in great slabs, making fire-fighting even more dangerous. Even when the fire was out, stripping what was left of the roof was worse than hazardous. But that is the next part of the story.

Meanwhile the photographs will show the damage better than words.

*Fig. 38. East side of barn on Saturday 26 April 1980*

*Fig. 39. North gable and west side*

*Fig. 40. South gable and west side*

*Fig. 41. South gable from west side*

*Fig. 42. South gable from inside – note hanging post II*

*Figs 43 & 44. East wall – bays 4, 3, 2, 1 from inside*

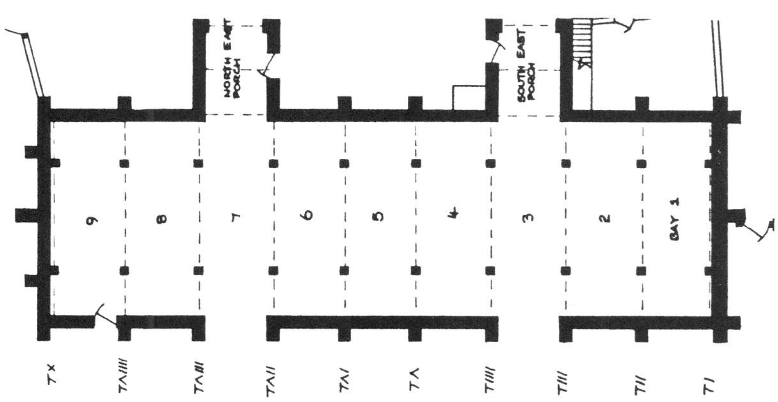

*Figs 45 & 46. East wall – bays 9, 8, 7, 6, 5 (part) from inside*

*Fig. 47. North gable from inside and homeless pigeon*

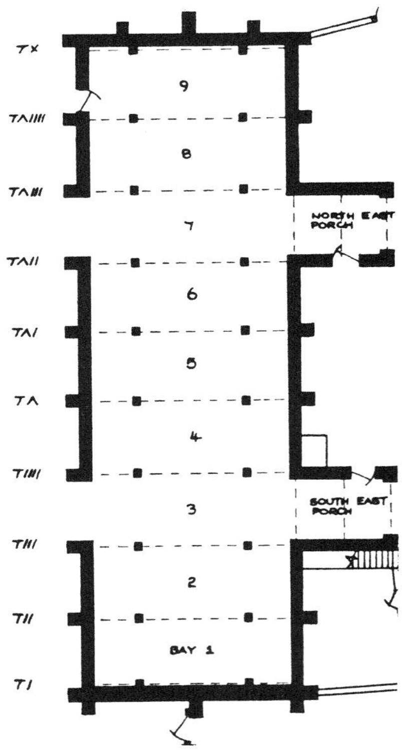

*Fig. 48. West wall – bays 1 and 2*

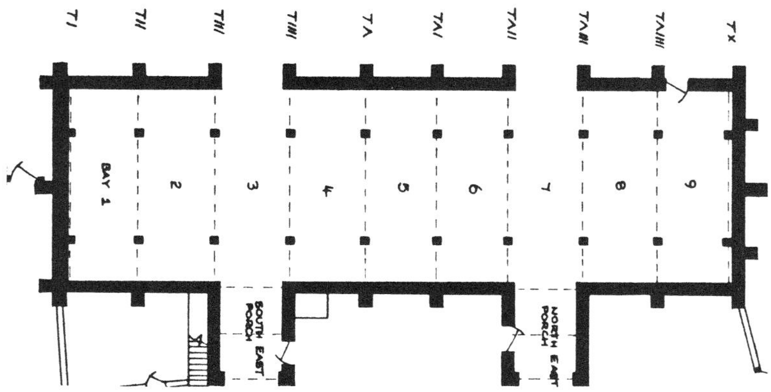

*Figs 49 & 50. West wall – bays 3, 4, 5, 6*

*Figs 51 & 52. West wall – bays 7, 8, 9*

*Fig. 53. South gable showing remains of trusses I and II*

*Fig. 54. East arcading and remains of trusses II to VII*

*Fig. 55. North-east porch showing north wall and remains of roof*

*Fig. 56. Looking into north-east porch from bay 7*

*Fig. 57. West arcading looking down barn from bay I – posts II to VIII*

*Fig. 58. Trusses V, VI, VII, VIII*

*Fig. 59. Trusses VI and VIII*

*Fig. 60. Trusses III to VIII looking north*

# Chapter 4

# EMERGENCY REPAIRS

The decision to rebuild Bredon was not easy for the National Trust. It was six months after the fire before confirmation was given by head office. In the meantime work mainly in the nature of emergency repairs had so far progressed that there was hardly an alternative.

Bredon, it should be remembered, was not only the last of four great barns burnt since 1962, it was also the first of five scheduled ancient monuments to be burnt in the last twelve years. Bredon was followed by the south transept of York Minster, next the Wren wing of Hampton Court, then a second National Trust property, Uppark, and, most recently, Windsor Castle. In such company Bredon may be the least distinguished of the buildings, but it raised all the questions about reconstruction and restoration, and it is certain that but for the experience of Bredon the trust would have to have thought much longer and harder about Uppark than they did, and their verdict might well have been the reverse. The other greater monuments obviously had to be rebuilt, by whatever means.

But Bredon also set a precedent for method. Practically complete authenticity of materials, construction and design is relatively easy to achieve in such a building. In more complicated buildings such a degree of authenticity would be difficult if not impossible. Thus, at York Minster the ribs for the vaulting had to be laminated instead of naturally swept oak, and the upper roof is a more efficient engineering design than the eighteenth century, and later altered, structure of before the fire – the medieval roof had already gone. At Hampton Court certain liberties also had to be taken with Wren's construction which too had been altered. But everything looks as it was; it is architecturally, if not structurally, authentic. Similarly, Uppark, where every vestige of original material saved from the fire has been reused, has achieved the same result.

There is another difference. Bredon had been surveyed in detail *before* it was burnt. We had not only my own sketches and details from the time spent at the barn with Horn's team in 1969, but Patricia Borne of Nottingham University had just completed a full archaeological survey of the building, and gave us copies of her immaculate drawings as soon as she heard of the fire. We were also inundated with photographs sent by others, but no amount of photographs could replace her survey. Thus we were prepared for the barn's reconstruction and saved the uncertainties and hazards of surveying what was left amongst the still smouldering debris and fallen timbers, as in the other buildings.

Thus there was no question that Bredon could be authentically rebuilt. Also considerable pressure was put on the trust, both locally and nationally. All the inhabitants of Bredon were of one mind and made their protest about the state of the barn and the trust's inaction very clear. In the wider field, Maurice Barley wrote an open letter to the Council for British Archaeology urging their trust members to write to the Director General or take whatever action they believed might be effective. I had also written a short report to the Architectural Press, published with their own forceful comments in the Architects Journal of 7th May.

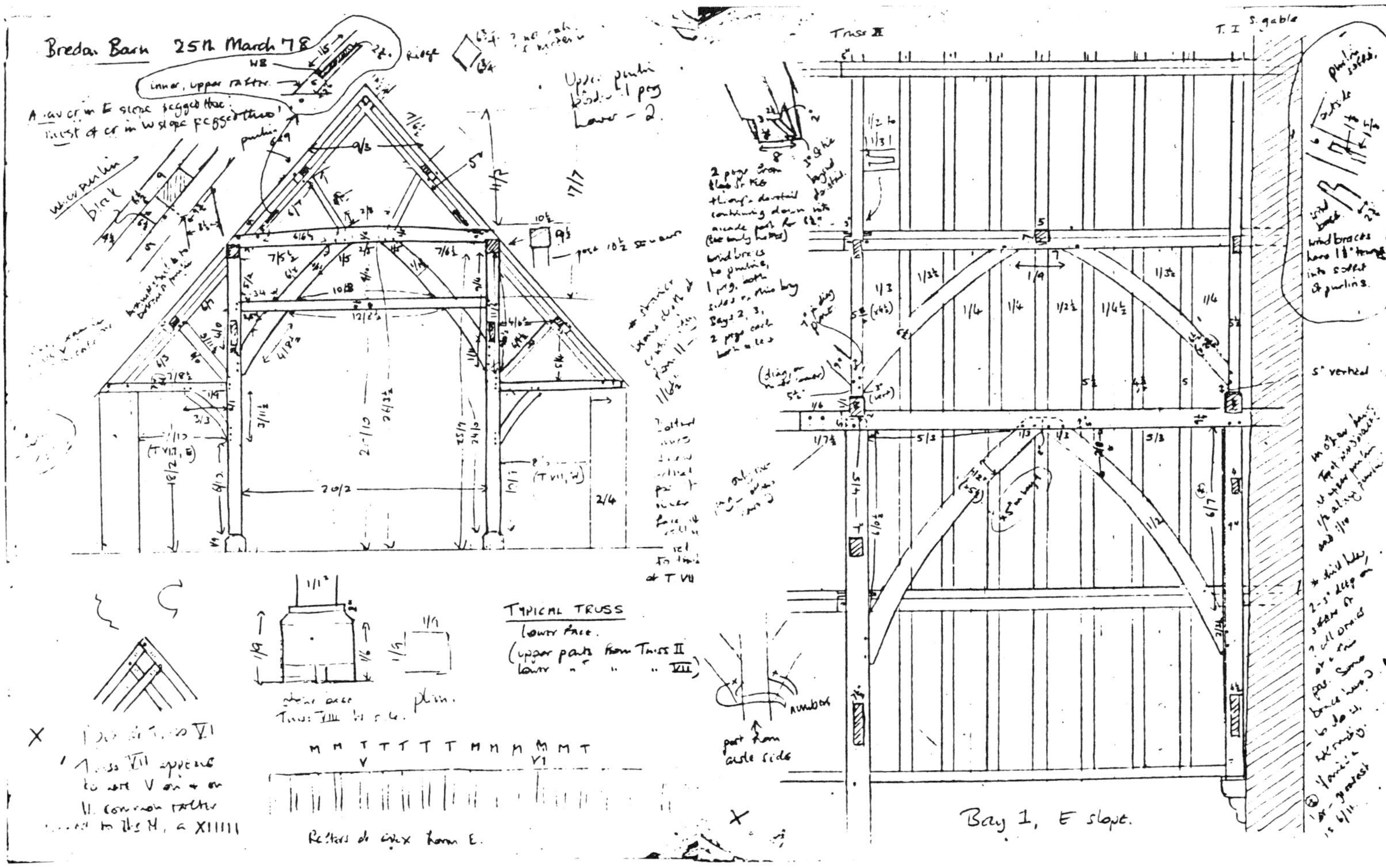


*Fig. 63. Examples of survey sheets by Patricia Borne*

Yet all the signs were that had not Finlinson appointed us a week after the fire (he had previously taken on another architect, a one-time assistant in our office, believing that he was an engineer!), head office would have done nothing, at any rate until they had debated and procrastinated until it would have been too late.

On 21st May, more than a month after the fire, we received the following letter. "The Director General has informed me that the Architectural Panel of the Trust would like to have a report from you before their meeting on 16th June on the proposals for restoration of this Barn. As I understand it, they would first like a factual statement on the structure and what remains of the timbering, followed by a discussion on the merits of full-scale reproductive restoration as opposed to alternate forms of reinstating the building as a whole and in this part of the paper remarks on the use of non-traditional materials". It was also suggested that partial restoration – bays 1 to 4 leaving the rest open as an archaeological site – should be considered.

Our report was already practically complete. I took it together with the photographic record to London on 12th June, four days before the architectural panel's meeting, and went over it with one of the head office members. He made little comment and gave nothing away about the views of the architectural panel. They obviously wanted no discussion with their architect.

Needless to say, our report had made no mention of "alternative forms" or "the use of non-traditional materials". The next demand was for the appointment of a consultant engineer. Fortunately, the regional office agreed to our recommendation of the firm of Anthony Hunt Associates. I knew Anthony Hunt by reputation and was confident that he would support our approach and would not automatically assume responsibility for everything structural. That the firm had a local office in Cirencester was a further point in favour of their appointment, and no objection could surely be made even if head office had any preferences of their own. Hunt and I met both at the barn and in our office and an appendix to our first report followed on 9th September, fully answering the Director General's requests. It read:

*"1. Alternative to Restoration*

The practicability of using other means and materials, has been examined not only by me but also by the firm of structural engineers, Anthony Hunt Associates. The conclusion is confirmation that exact restoration in original materials, using every piece of the existing timbers that can be repaired, is the only right and practicable course. Even on grounds of cost, the engineers state that the substitution of a steel-frame for the oak would show "only a marginal saving". The possibility of a reinforced concrete frame is not even mentioned but it was discussed and discarded because it would be considerably more expensive. So also were all the other materials and structural methods available. Briefly, this is because the problems of marrying modern materials and techniques into traditional construction were found to be insuperable.

All that the engineers could recommend therefore was a temporary structure under which it would be possible to carry out proper restoration over a period of years. Against that, its cost of £34,000 could be better spent on the permanent work. Partial restoration was the third alternative, leaving the three northernmost bays unrestored and unroofed for a indefinite period and without commitment to complete restoration. This however would necessitate construction of a permanent wall or partition across the building at whatever truss it was decided to terminate the timber-frame. Its construction, requiring foundations, a means of joining it to the stone flank walls of the barn, as well as the frame, and provision of a permanently secure and weathertight verge for the roof would be as full of problems as the first ideas. A traditionally-built stone wall, a replica of the original that still stands repaired and ready to resume its proper function, would be practicable but would waste the latter, and be much more expensive to construct than proper restoration of the timberwork and its termination in the present wall.

*2. Current Work*

The appearance of the barn today is in startling contrast to what I had to describe and illustrate in June. All the destroyed lintols have been replaced in oak, obtained at exceptionally low cost from a little known supplier in Devon. The damaged masonry over the lintols and, of greatest structural importance, at the apex of the south gable has been repaired. Stone repairs are still going on, including the verges, the destroyed corbel stones of truss I, and soon the capping of the medieval chimney.

All the collapsed stone that could be rescued is being reused, but new stone has had to be obtained for five of the post plinths, for the corbel and chimney – the last not damaged by the fire but in urgent need of attention through age. Incidentally, the scaffolders erected additional scaffolding, hung from the main scaffold, without extra cost to make the chimney accessible.

The sheeted scaffold is complete and protects trusses III, IIII, V (bays 3–4) and the inner end of the south-east porch. These trusses will not have to be dismantled. The covered area is adequate for repairing the other trusses, if not for storing them. All the more seriously damaged trusses, I, II, VI and VII, have been taken down and their components are lying in an organised way in the barn or being made ready for repair. The charred surface is about a quarter to half an inch deep on all the burnt surfaces and this is being scraped off to reveal a very beautiful darkened oak surface.

Truss VI is in process of being repaired. It will be completed in about three weeks. The operations are being closely watched and will give a reliable basis for the labour costs of subsequent trusses.

*3. Preparation for Restoration*

On the office side, the specification and details of all the new timber have been issued and the contractor is obtaining three competitive tenders, to be submitted by each as a lump sum. As soon as these have been received the lowest estimate should be accepted without varying the quantity. The competition for this order is intense, both because there is a momentary lull in the rising price of oak and because the suppliers see this as a prestige order. One supplier has stated that his cubic foot price, which is about 25 per cent lower than in a quotation for another contract a few months earlier, will hold only if he is given the complete order. There are three potential suppliers all of whom have quoted an exceptionally low cubic-foot rate.

The estimate of the Forestry Commission for chestnut poles for the rafters, mentioned in the earlier report, is also remarkably favourable; there would be little saving even by the use of sawn softwood which would be technically and aesthetically wrong for this building. Other estimates are included in the Quantity Surveyor's "Budget Guide" of 5th September 1980.

All the planning and documentary work is now complete, including the design of each truss destroyed in the fire – T VIII, VIIII and X –, the inner truss of the north-east porch, and the repair joints and timber replacements of the other trusses and longitudinal bays. The stone details have also been completed.

The outstanding question is still the method of dealing with the roof. An estimate for steel sheeting instead of Cotswold slates shows that £70,000 would be saved on current prices by adopting this expedient. It is the only acceptable alternative if funds cannot be made available for total restoration under the present contract. The danger is that the temporary roof may become permanent, as it will always be cheaper to repair than to replace it with Cotswold slates – especially as Cotswold slates become scarcer and more expensive.

However a way out of this predicament has become a possibility since the date of my previous report. The reopening of a quarry at Filkins, near Burford, on land owned by Sir John Cripps, is being considered. This is the subject of a report by Mr R G Franklin, RIBA, which was sent to the trust and other non-commercial bodies. I am told by him that an order from the trust for the slates needed for Bredon alone would be enough to guarantee the economic viability of his and Sir John Cripps' project. An immediate commitment would also obviate expenditure on temporary roofing as slates could be delivered by the beginning of 1982, when the whole of the structural work should have been completed. How this would compare in cost with Broadbents' estimate of £80,000 is not known; it would certainly prevent the vandalising of other buildings.

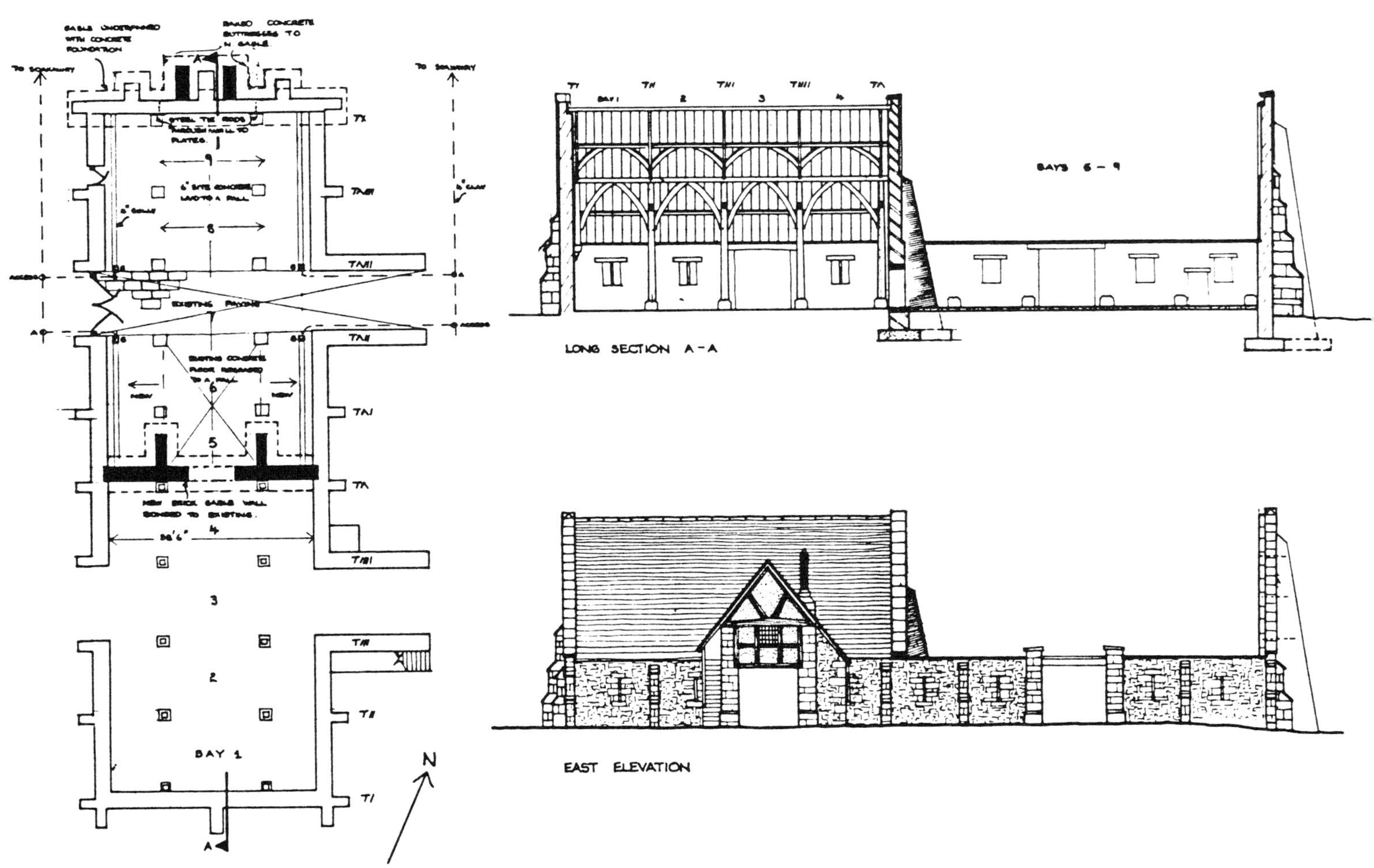


*Fig. 64. Plan, section and elevation of scheme as suggested by National Trust*

To summarise, in the space of less than five months since the fire, the prospect for the future of Bredon barn has been transformed. Instead of a hopeless ruin the scene is now of purposeful reconstruction. Everything has been done to maintain progress with the contractor, Spicers, taken on by the Trust even before I was appointed. Their men have gained the kind of experience that will enable the contract to go through to total restoration without a hitch. I would only emphasise that continuity is absolutely vital."

We drew the castrated scheme and it was never heard of again. The temporary structure designed by Hunt was in the form of a hyperbolic paraboloid, like a vast tent slung in a double curve. This scheme too simply disappeared, and the trust were left with little choice.

Through all this time, we had known that if work were stopped and Spicers' men taken off the site, the chances of its restarting would have been remote indeed. We were always supported by the regional office who never questioned our decisions as to the next "emergency" repair or our instructions to the contractor to put work in hand or order materials. Of course with our Quantity Surveyor, P V Firminger, to estimate every item beforehand and inform the trust, expenditure never got out of control.

First, the hanging post II was propped as soon as it was relatively safe to enter the barn. This was on the 28th April. Stripping the roof took much longer and entering the barn was not safe until all the slates, likely to fall in great chunks owing to their mortar bedding, had been taken off and stacked. Broadbents, the roofing specialists, were called in and with the use of their "pussycat" for sliding the slates to the ground completed the job including their methodical stacking by end of May.

*Fig. 65. Post II showing first of emergency repairs*

*Fig. 66. Chimney capping before repair*

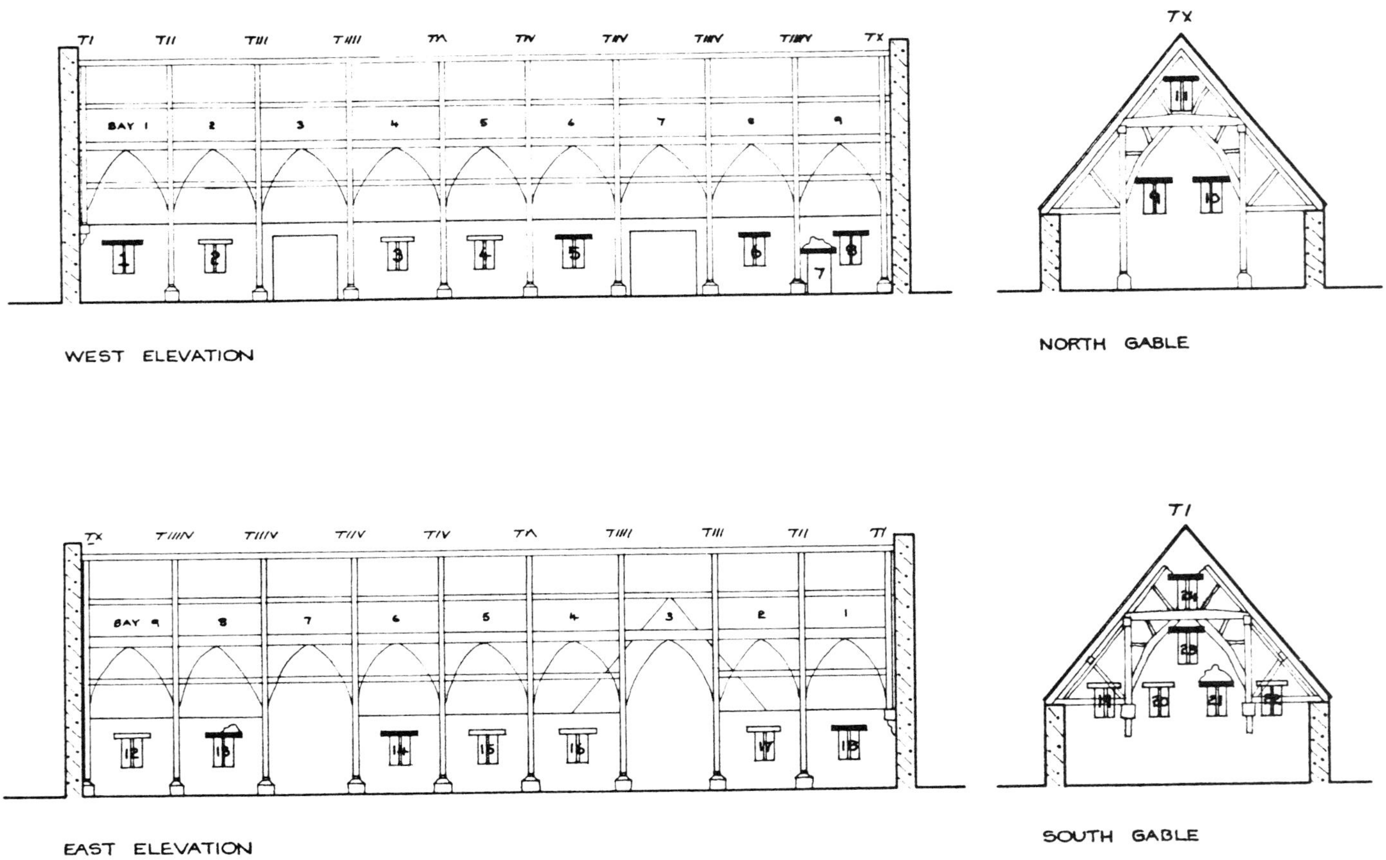


*Fig. 67. Replacement of window lintols (black) in west elevation, east elevation, north gable and south gable*

Next, the crumbling masonry above the burnt out lintols had to be restrained and the lintols replaced, especially in the south gable wall. This meant a survey and schedule for Spicers to be able to order oak, issued on 8th May. The size of the lintols in the south gable was 7′6" long by 2′3" in wall depth, in three separate sections, each 9" wide. Their placing was possible only after all the loose stone above them had been taken out. But because their splayed reveals narrowed to 6" on the outside none of the external masonry was touched. A tower scaffold was moved from one window to the next. All lintols were in and the walls rebuilt over them by the end of June.

Other stone repairs followed. New stone could be obtained almost immediately from Ben Davis of Worcester, and Spicers had masons for everything except carving. New corbels for the south gable were delivered in early August, the main stone of the west corbel having the year "1980" incised on it. Plinth stones for post VII and a pair for posts VIII were also delivered. Both of the side walls of the north-east porch at their junction with the barn had to be rebuilt, not only had they been damaged by falling timbers but at some time the top quoins had been replaced with brickwork. New quoins were obtained and the corners rebuilt. The chimney, as noted, was not damaged by the fire but urgently needed renewal of its eight dwarf columns of the side outlets below its dunce's cap.

During all this time, Russell Jones was not only detailing these repairs but surveying the whole of the timberwork. We issued a survey of the state of every truss and aisle framing, accompanied by proposals for repair and replacement (our typical S & R drawings – numbered 1 to 16) on 24th June. All the repairs and replacements shown had to be checked and often revised when the trusses had been dismantled and each timber could be individually examined on the floor.

Meanwhile the farmer had removed his machinery from bay 7. He had also removed to the west side of the barn the fallen tiebeam of truss VII. He had tried to axe it into two halves but had only succeeded in chipping out two wedge-shaped cuts top and bottom. It was beyond reuse but had to be kept as a template for its replacement. He was stopped from removing any more timbers.

Lastly, the north gable was tackled and all masonry repairs, except replacing damaged stones of the oversailing head of the side walls not done until much later, were completed by mid-September.

*Fig. 68. Bays 7, 8, 9 and repaired north gable*

Truss I was the first to be taken down, not because it would be the first to be re-erected, but to make the gable wall accessible for repairing its stonework. Truss VIII followed, and then trusses VII and VI, the reverse order of their eventual re-erection. At the same time, the longitudinal framing had to be removed between one truss and the next. The work was dangerous as the larger timbers could weigh anything up to half a ton and everything from releasing the joints to lifting them had to be done from tower scaffolding.

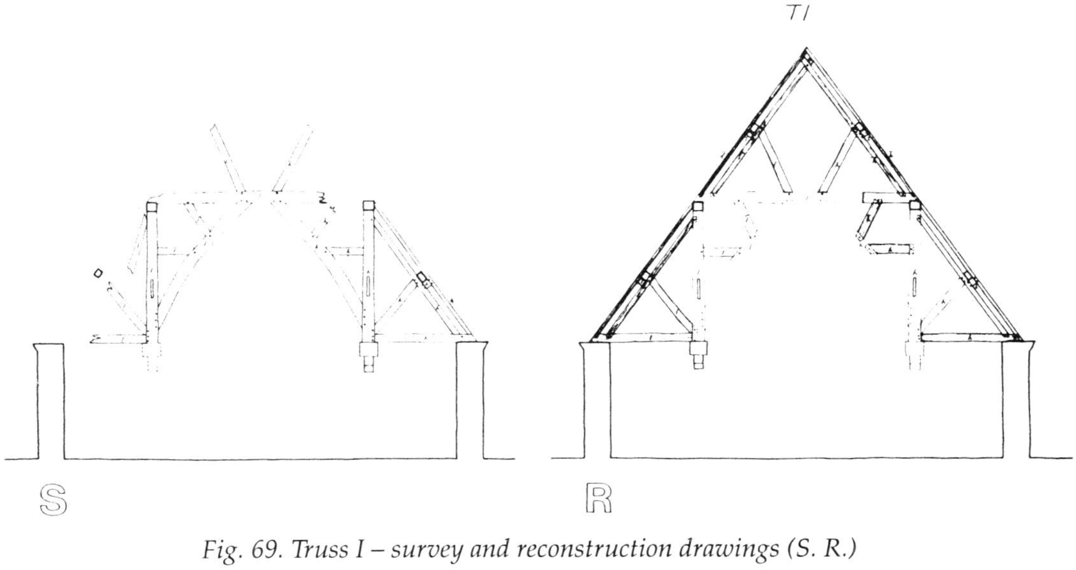


*Fig. 69. Truss I – survey and reconstruction drawings (S. R.)*

*Fig. 70. Typical number plaque*

*Fig. 71. Fallen timbers numbered*

Every timber as it was taken down or where it lay on the floor had to be numbered. The original system with its orderliness and clarity was of course followed, but the numbers were made more clear by a numbered 4" square plywood plaque screwed to the "upper" face of each timber. Only the components of truss I, erected against the gable wall, were numbered on their lower face.

*Fig. 72. Scaffolding - sheeting in progress*

*Figs 73 & 74. Top and bottom faces of burnt arcade-plate*

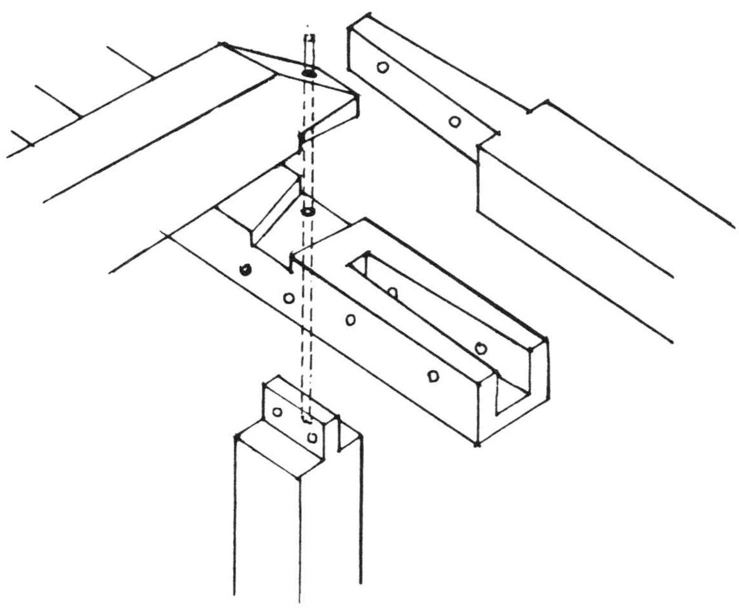

*Fig. 75. Diagram explaining photographs of arcade-plate*

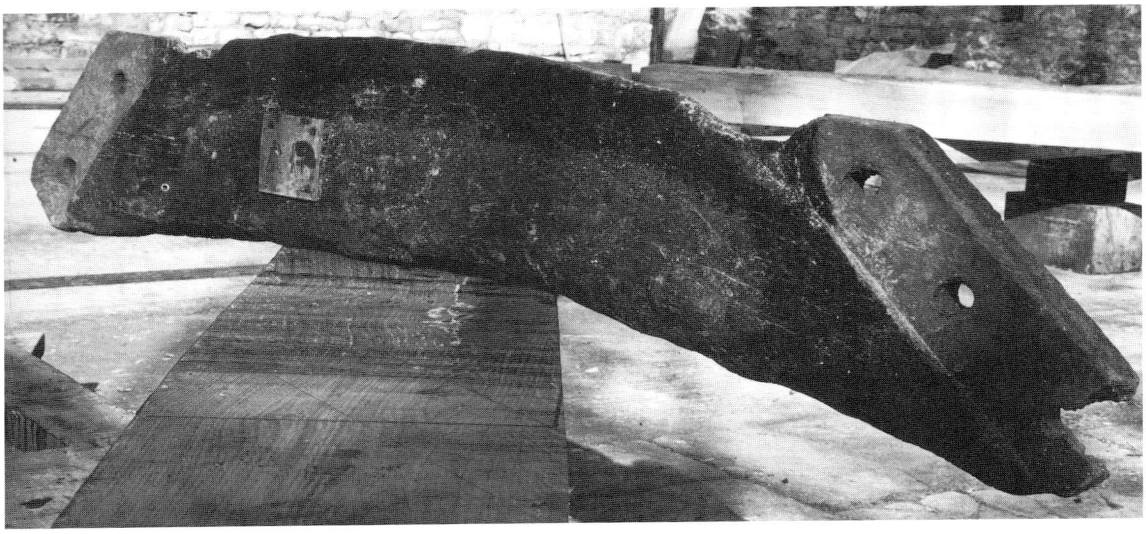

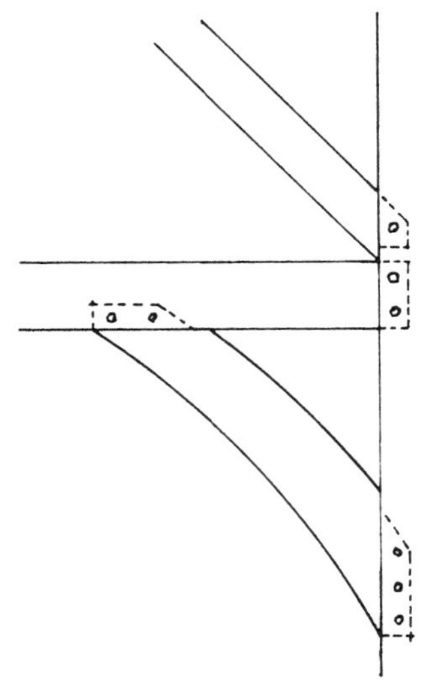

*Fig. 76. Aisle brace from truss VI*

*Fig. 77. Diagram showing aisle brace as fitted*

The other major work was scaffolding. Ideally we should have enclosed the entire building, but the cost of this would have been prohibitive. Instead we designed a scheme for protecting the main cross-frames only, permitting them to be re-erected up to tiebeam level. The upper roof would then be erected in the open after dismantling the scaffolding. This would have left the gable walls exposed throughout. In taking this risk we were assured by the precedent of the Beaulieu barn with its solitary gable standing after four hundred years without support or protection. Perhaps more to be afraid of was the possibility of the trust seeking the advice of an engineer other than Anthony Hunt who would almost certainly have demanded shoring and then, almost as certainly, underpinning and new foundations. This too would also have damned reconstruction before it had started.

Even the reduced scheme according to the estimates obtained by Spicers would have cost more than £30,000 which we considered too much to present to the trust. Our finally reduced scheme, covering only the undamaged bays 3 and 4, would cost £10,000. This was accepted by regional office and the scaffolding was erected and sheeted by the end of June.

The only other preparatory work was concreting the floors of bays 8 and 9, as they would be open to the weather. The porches and bays 3 and 7, the threshing bays, were already paved and bays 1, 2, 4 and 5 had previously been concreted.

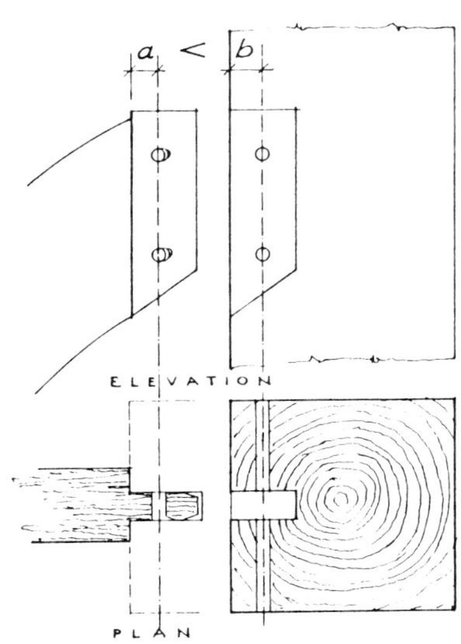


*Fig. 78. Draw-boring – photograph shows gouged-out peg-holes permitting entry of pegs; diagram showing method of drilling peg-holes so that tenon is tightened into mortice*

*Fig. 79. The only surviving pieces of the ridge-pole with upper joint (female)*

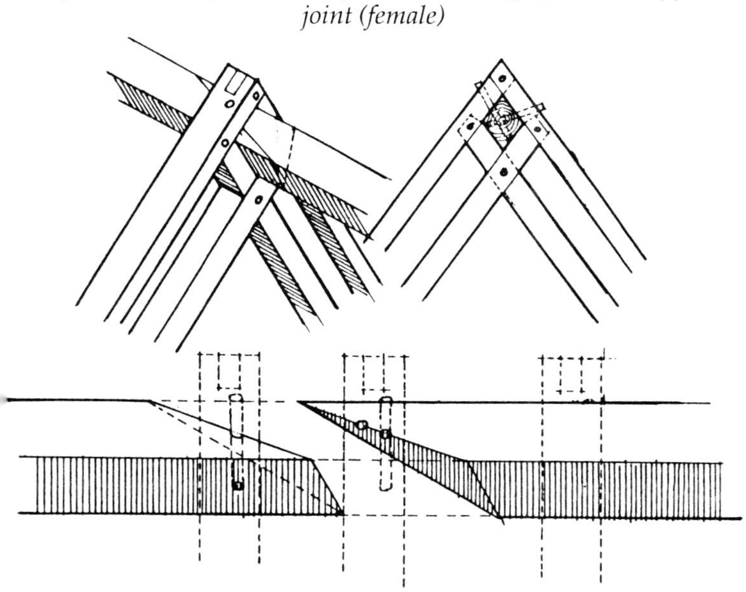

*Fig. 80. Diagram of ridge-hole as fitted at each truss and male and female joints*

The burnt timbers revealed details of the original construction only visible when they had been dismantled. Figures 67 and 68 show the arcade-plate at truss VI, west side. The top photograph is of the upper side with its dovetail for the tiebeam and to the right of that the 18" long, tapered bridle mortice. This joint is unique in our experience, and as elegant as any medieval carpentry detail we have discovered. The designer-carpenter must have visualised the load from one arcade-plate to the next being transferred gradually, less at the inner end and increasing to its greatest at the point where the tenon of the adjoining member had to take the full sheer load. The tapered form also requires the greatest accuracy in cutting both parts of the joint.

Another concealed detail was the method of fitting apparently contiguous members, as the aisle-tie and aisle-strut where they enter the outside face of the main posts. The tenons are cut back at the shoulder so that there are two mortices, instead of a single elongated one, to house each tenon separately (figs 69 & 70). This was general for such swept braces but not always uniform. It probably depended on the size of the branches before axing or adzing them, the wider ones being shouldered, the narrower used full. When it came to the new trusses, VIII, VIIII and X, Dauncey and his carpenters were left to make their own judgement, as their forebears, but replacements of those braces too badly burnt for reuse had to follow the original.

The aisle tiebeam brace of T VII had been badly burnt at both ends. The fire had licked up the post and along the soffit of the tiebeam, burning away even the lower end of the brace tenon within the post, so that its original three peg holes were reduced to two. The brace was reused as it was, even though the tiebeam had to be renewed and the post repaired with a long patch.

The ridgepole had been almost completely destroyed throughout the whole length of the roof. One piece found showed the design of the scarf-joint and its manner of pegging. The photograph shows the "upper", or south, section of the joint. The peg holes took pegs inserted through the rafters. They are not in the same alignment on either side of the ridge, nor is their direction at the same angle; one is nearly vertical, the other almost horizontal. Thus they avoid each other when driven in.

Draw-boring (fig. 71) was found in every dismantled mortice-and-tenon joint at Bredon. This is the method of first drilling the peg-hole on the outer wall of the morticed member and marking the tenon with the bit or auger, but not penetrating it. The tenoned component is then withdrawn and its peg-hole drilled slightly closer to the shoulder of the morticed member. The original peg-hole is then completed by drilling the opposite wall in the same alignment. It may be seen how the peg-hole of the withdrawn tenon has been gouged to ease penetration of the pointed peg. Such pegs have to be hammered in with considerable force, crushing and distorting the peg in the process, as seen when it has been withdrawn – again by vigorous hammering. To drill pegs out would obviously destroy the purpose and effect of draw-boring which is to tighten the joint against the shoulder of the tenon to its maximum.

With all timbers down and numbered there was nothing else to be done. Emergency repairs had run out and the men were reduced to wire-brushing the timbers to remove their charring. Chambers had been to our office fearing an even blacker future, and Dauncey was in despair at not being allowed to start any timber repairs. But at the site meeting on 28th July all this was changed when Finlinson announced that truss VI should be repaired as a basis for costing the whole job and to prove that it could be done.

# Chapter 5

# RECONSTRUCTION

There was still no confirmation from head office, but timber was ordered beyond the limited needs of truss VI. It would have been too uneconomic to do otherwise. Venables of Stafford had already offered the most favourable estimate on the basis of our timber schedules, on condition that they would be given the whole contract, and it could hardly be expected of them to select, convert and deliver a few varied pieces when the cost for a full load would have been scarcely greater.

*Fig. 81. Bay 3 as workshop – undamaged trusses III and IIII*

But there was plenty of other work before deliveries could begin. First Spicers had decided that, instead of repairing the timbers and re-fabricating the trusses at their yard in Worcester and delivering them to the barn ready for erection, everything should be done on the site. This would save on handling and transporting heavy timbers, but it greatly complicated Dauncey's task. First, the floor had to be cleared to complete concreting in bays 1 and 2 where all the dismantled timbers had been stacked. The covered bays 3 and 4 had to become the workshop. The remaining bays would be used for laying out each truss in succession for re-framing before being taken apart again to await its re-erection. Originally, of course, the trusses would have been finally pegged and reared on the spot, but Spicers had also decided against rearing. Thus, space had to be reserved for the frames both before and after their repair, with the latter requiring much more of it. In addition to the main trusses, there were the principal rafters of the top roof, the three trusses of the north-east porch and the whole of the arcade framing.

The available area was only that between the barn and the garden wall. The fence and gate at the north end were moved northward as far as the slope of the ground to the river permitted, but this only made room for one of the portacabins, the messroom. At the other end there had to be turning space for the timber lorries and for

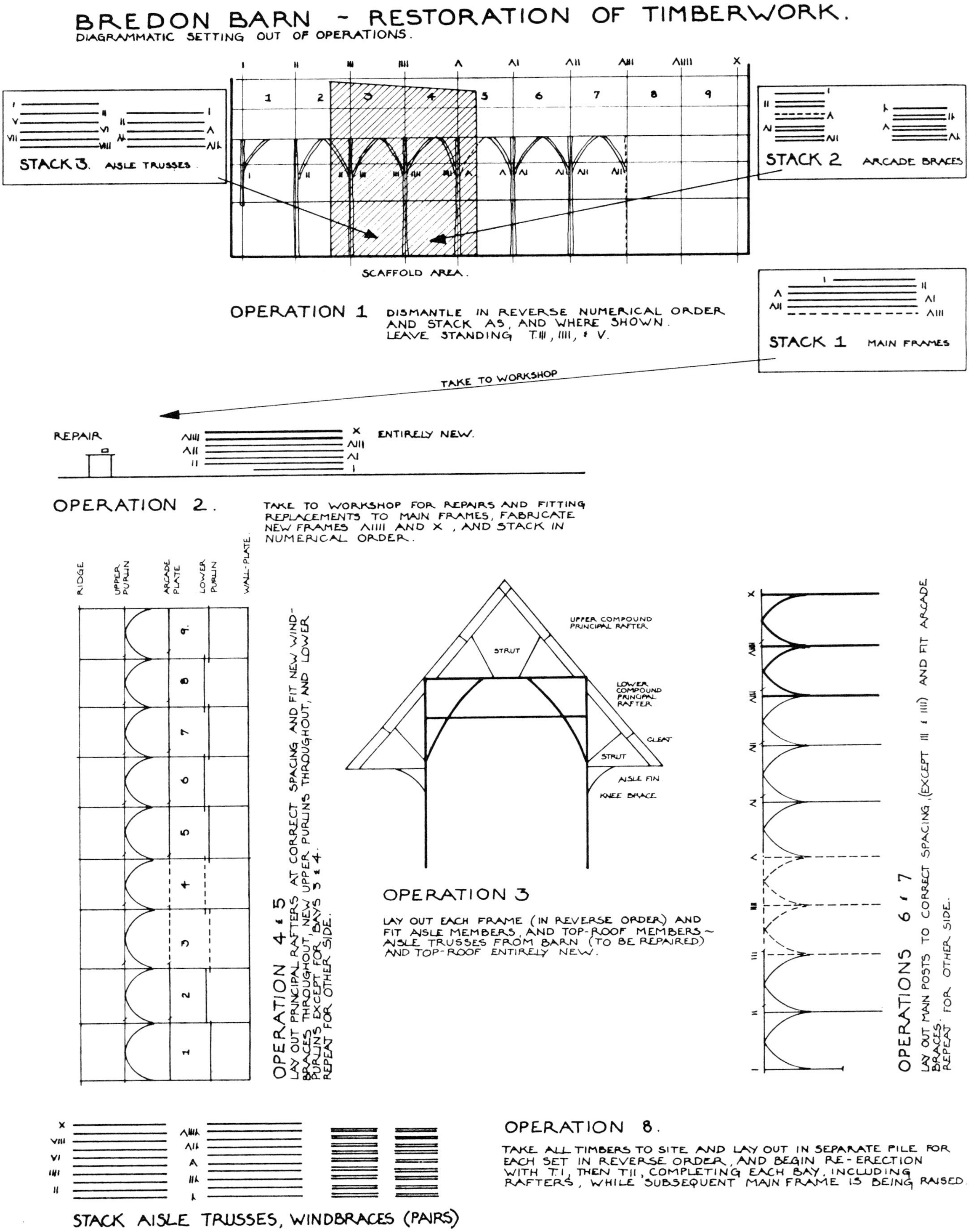


*Fig. 82. Instructions for organising timbers*

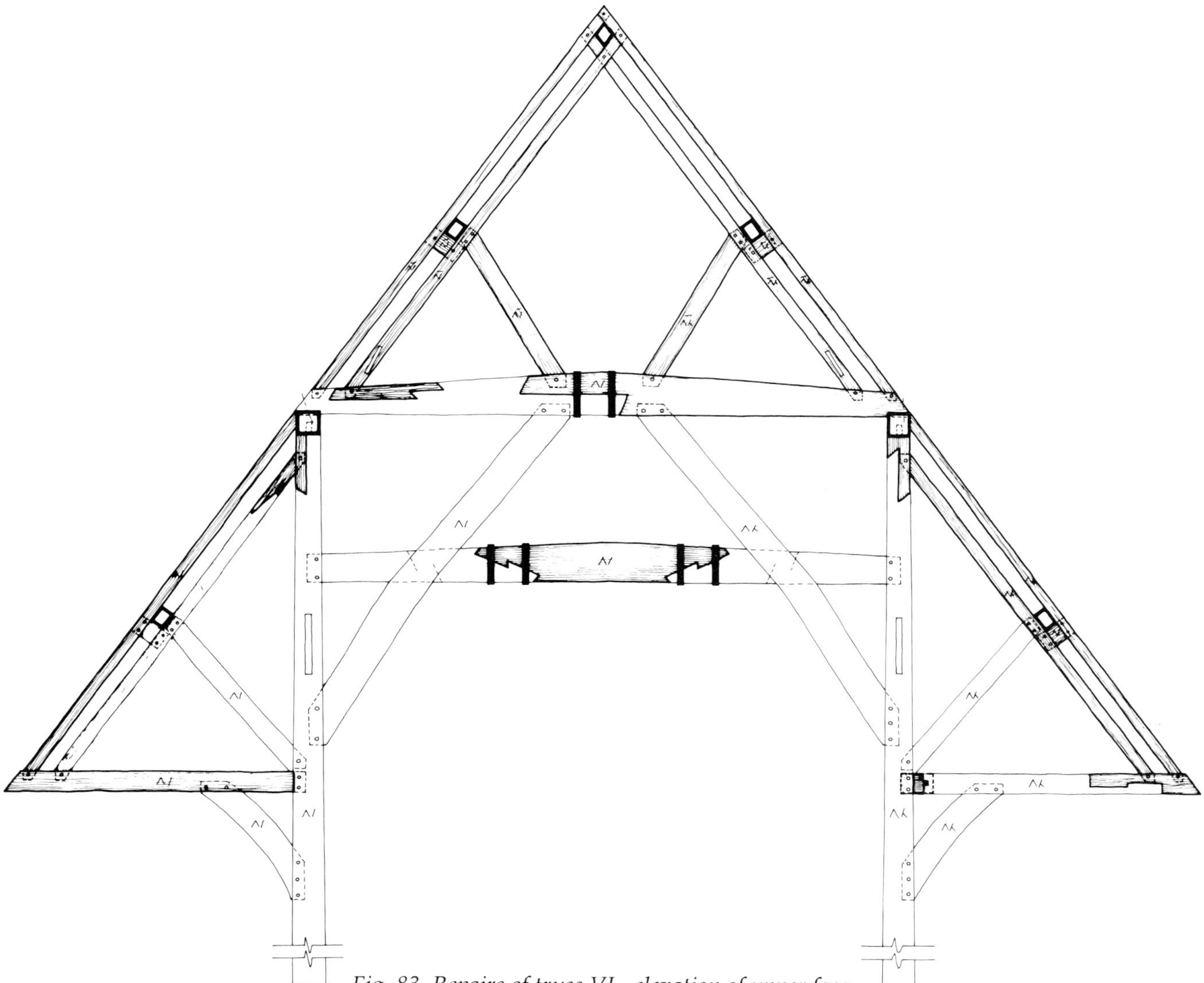

*Fig. 83. Repairs of truss VI - elevation of upper face*

unloading, and for the other portacabin, the site office. In between could be used for equipment, including the mobile crane with room to manoeuvre and all the stacking of ever-increasing quantities of timber. The method of stacking had to follow, at least in principle, the drawing issued much earlier on the presumption that Spicer's yard would be available (fig. 75).

How Dauncey managed the continuous process of moving every component of every frame for the next operation without, so far as I know, ever having to dig to the bottom of the pile, or forgetting where to look, speaks volumes for his organising ability and memory. The other side of the barn belonged to the farmer and could only be used for storing some of the salvaged slates.

*Fig. 84. Carpenter's setting out lines for jointing new timber*

Truss VI was of course the first to be taken into the barn workshop (bays 3 and 4) and now that each component would be examined on the bench its repair joints would be finally designed, not without Dauncey's contribution. The repaired members would have to stand up to, not only the stresses and strains within the completed structure, but also the far greater ones of handling, hoisting and hanging from a crane.

Both the main beams of truss VI were badly damaged. The tiebeam was burnt at both ends, most severely its

*Fig. 85. New head to post VI*

*Fig. 86. New end to aisle-tie, truss VI*

eastern half, so that the repair had to be more than half the length of the original. The scarfed-in patch at the other end might appear to weaken the beam, but since it is in the upper half of the beam's depth it is in compression. The under-squinted butts therefore cannot loosen through shrinkage or any other cause. The patch also has to be of such accuracy that it must be hammered into place. Again it is finally secured by skew-pegging.

*Fig. 87. Fitting new section into tiebeam, truss VI*

*Fig. 88. Truss VI repaired*

The strainer beam, oddly, was burnt at midspan. Here it had burned right through (both beams as they appeared after the fire are best seen on fig. 52, Chapter 3). Its repair, in contrast with that of the tiebeam, consisted of a large replica of the burnt original, scarfed into its two ends. Both of the beams had to be strapped. The design of the straps was hardly less important than the joints. The first ones delivered by Spicers were welded at the angles. They had to be changed for continuous steel, bent instead of welded. The slightly rounded corners could then be recessed into notches, cut into the timbers at their corners, preventing any tendency to creep. Tightening was only "thumb-tight" but could at any time be adjusted. The post-heads and the tiebeam of the east aisle also needed new ends.

Truss VI was completed on 3rd October. Its repair had taken two months and proved that the job could be done. Whether this made any difference to deliberations at head office or not, confirmation that Bredon barn would be fully reconstructed at last came through on 20th October.

There was little noticeable difference in the sitework, but Spicers' contract could now be signed, and an estimate by C.F. Miles, a Cotswold roofer, for £60,000 for battening and slating the roof – £20,000 less that Broadbents' – could be accepted, as also Venables' estimate for the timber.

Truss VI had already been dismantled and its components once again stacked to await its eventual re-erection, and truss VII and the arcade framing of bay 6 had been laid out on the barn floor for the repair of their timbers. This would test Venables' resources, for such members as a full-length tie-beam, almost full-length posts and, most testing of all, a pair of mirror-image arcade braces would be needed.

Our specification was no more than the interpretation of original timbers as found in any medieval building. The carpenter of that time used the *smallest* tree that would yield the required finished member. Even for a structure as monumental as Bredon's he would have had to fell, transport and convert no oak of more than 60 to 80 years' average rate of growth. Each such tree could be halved to make a pair of tiebeams, or squared for a post, or again halved for a pair of the large mirror-image swept braces – probably obtained from coppiced oak.

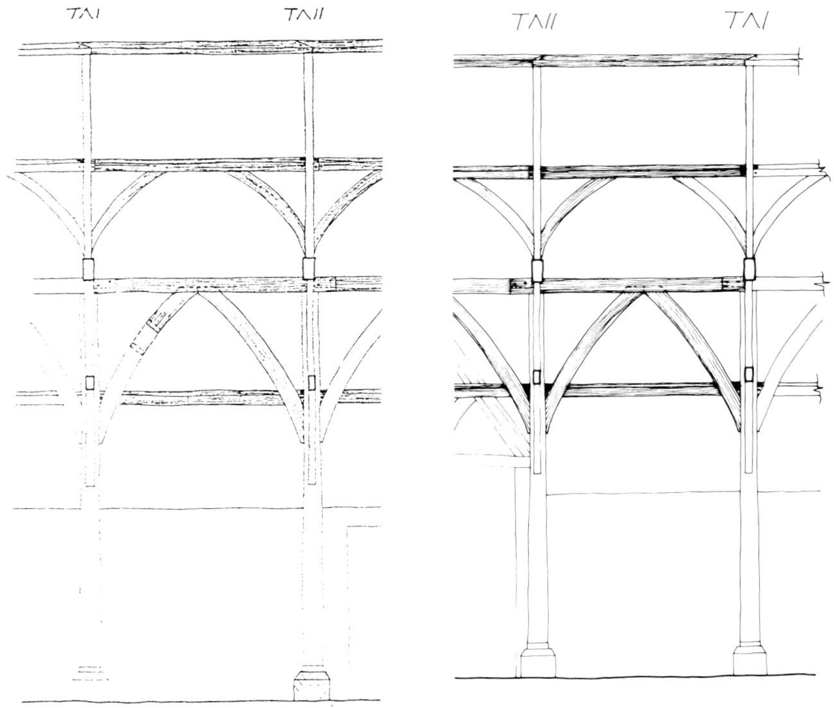


*Fig. 89. Repairs for bay VI, west arcade and east arcade*

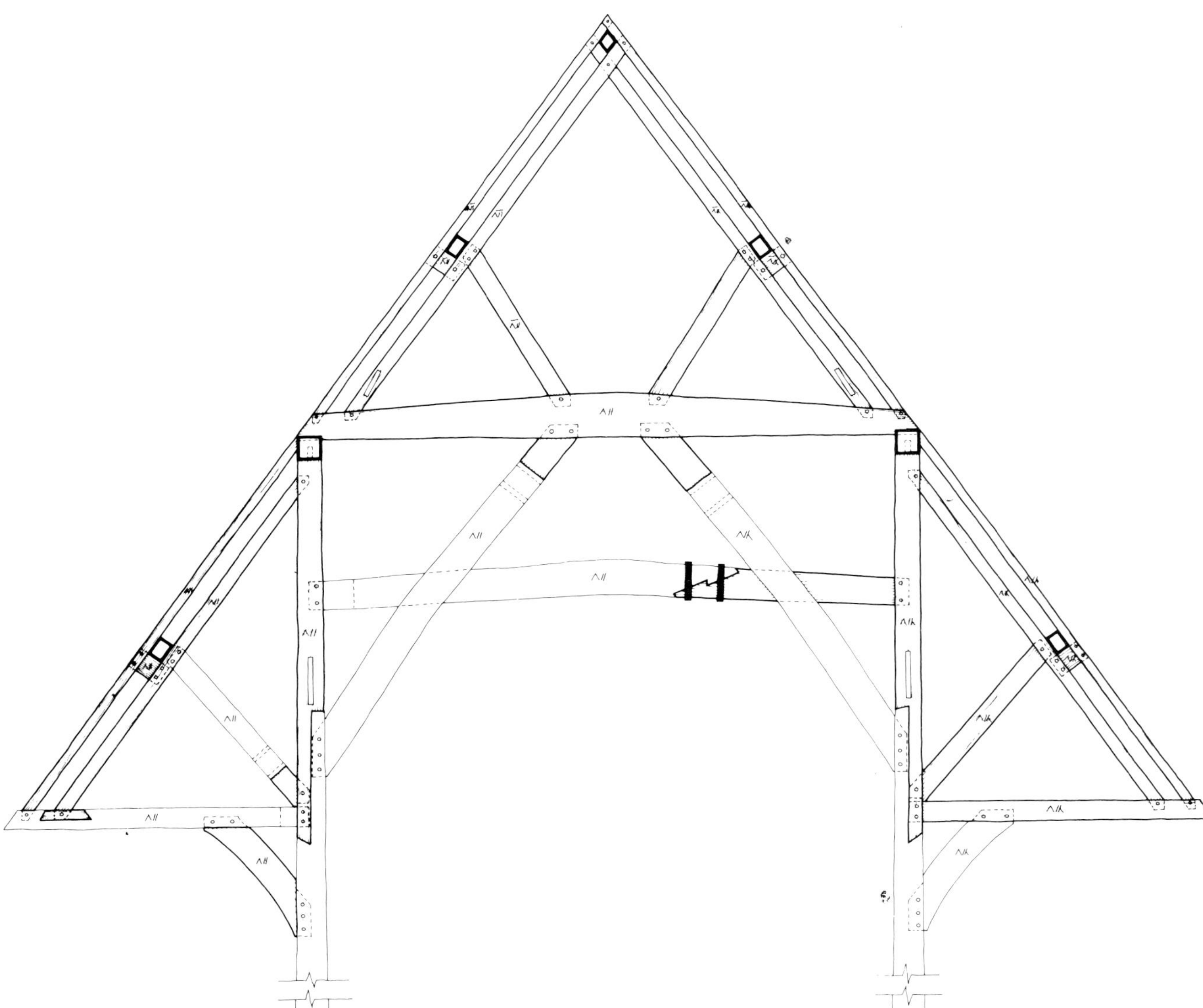

*Fig. 90. Repairs for truss VII*

*Fig. 91. Rafters as delivered for final conversion*

*Fig. 92. Bandsaw ridge on post VIII*

By contrast, the timber merchant of today, with mechanised transport and powered bandsaws, requires trees of 3 to 4 feet diameter, about hundred and fifty to two hundred years old. From such logs, sawn through-and-through, he might obtain half a dozen posts and as many beams, but only one post from the heart of the log and a pair of beams would be acceptable. A swept brace, unless the log were almost too bent to mount on the bandsaw, would be impossible to produce, except by shaping the slab to a template after it had emerged from the saw. This however would be to disregard the natural grain of the oak – the very essence of timber-framing.

Every timber as it was delivered, often before it was taken off the lorry, was examined by Russell or me. In halved members the heart-shake had to be visible throughout the timber's length. In boxed-heart posts, arcade-plates and other square components the heart had to be visible near the centre of both of the sawn ends. In swept members the curve had to follow the grain with the heart again visible throughout the length, and each pair a mirror-image. Many complete loads were sent back, the arcade-braces three times before an acceptable pair arrived.

Even rafters, when their time came had to be rejected. They should have come from 'flitterns', oak poles normally too young to fell. They could be sawn along the heart and squared along their sides with the axe, giving rafters that would lie flat on the ridge, purlins and plates but wander slightly in the other dimension. The first rafters delivered had been sawn straight on all four sides from a log that would give four rafters. Eventually Venables provided heart sawn poles to be finished on the site. Dauncey and his carpenters were certainly capable of axing down the sides, but this was too slow for Spicers. Chalk lines had to be drawn along the grain, followed by the chainsaw. The carpenters were then put to work with chisels to eliminate the saw marks, taking twice as long as would have axing at the beginning. This was too much, and we accepted the saw marks as preferable to any further simulation – or dissimulation. They would honestly date the rafters just as all other new timbers were dated by the unmistakeable ridges of the bandsaw. But it was far to early to raise any rafters and they too had to be found space for stacking for an indefinite period.

All the trusses, VI to X and their aisle-framing, and trusses XI, XII and XIII of the north-east porch were completed and stowed away again by the end of March 1981. The porch trusses could be kept re-fabricated in the barn. Then was the time to start at the south end with truss I, already dismantled and stored for the best part of a year. The plan agreed, as tersely recorded in the job book on 16 April, reads:

*Fig. 93. New truss VIII laid out for framing*

*Fig. 94. Repaired truss XII, north-east porch*

1. Complete *all* repairs
2. Repair trusses I and II and bays 1 and 2
3. Take down scaffolding and sheeting of bays 3 and 4, and erect new lean-to against north gable wall as workshop
4. Erect trusses in numerical sequence, I to X
5. Begin top roof and raftering at south end

Dauncey was driven almost to despair, "Here we have been working for over a year, and nothing at all to show. How had those old chaps done it in a year – walls, timbers, roof, everything, and what must all these important people from London think when they come to see what we're doing?" And there were still months of repairs to be done before there would be anything to show. The timbers of truss I, the first to be dismantled almost a year ago, had to be found – only Dauncey knew where – and laid out for repair. They were far more burnt than Russell had been able to see when

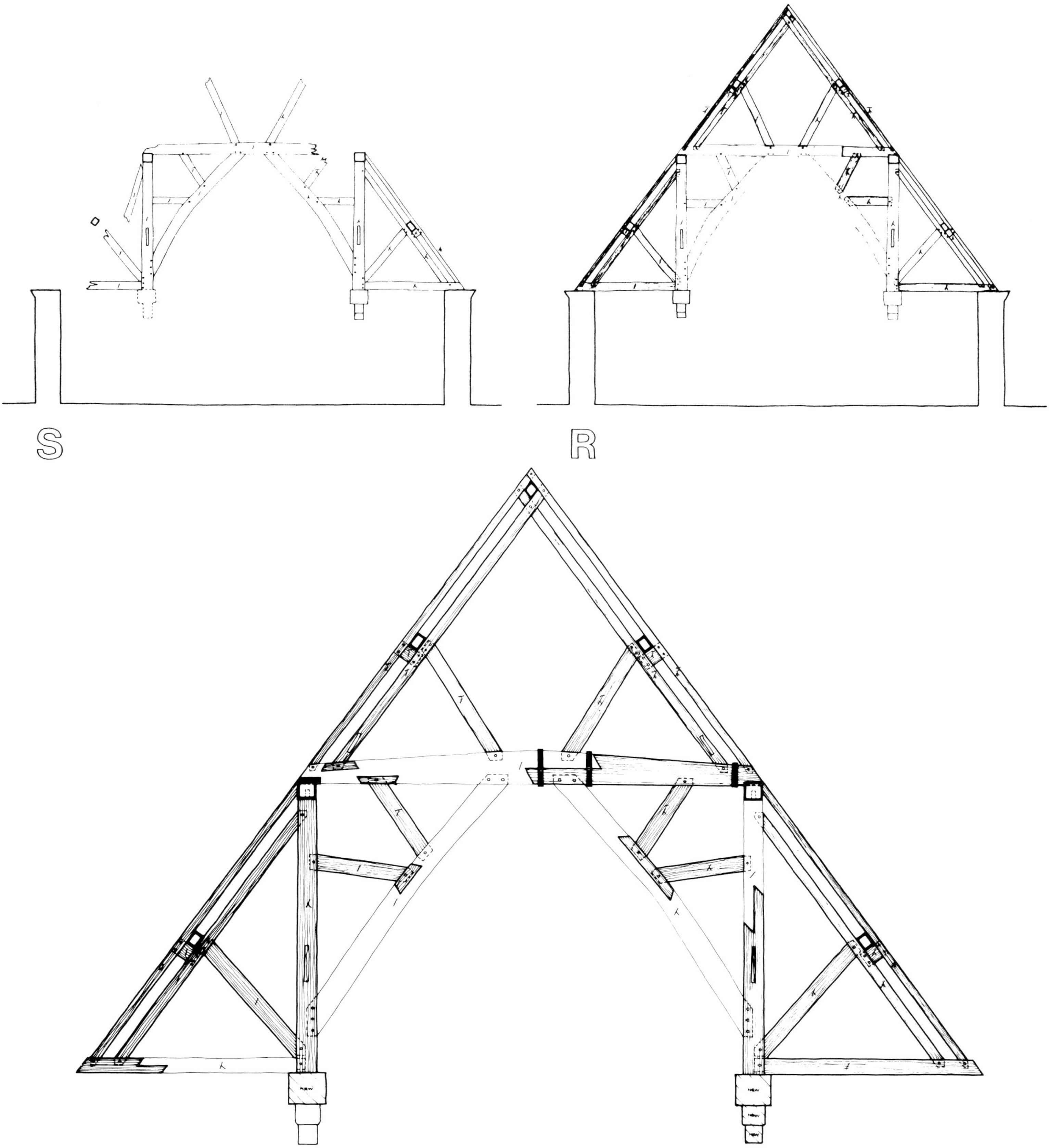


*Fig. 95. Truss I: S as surviving after fire, R repairs as intended, and bottom as finally repaired. Compare with fig. 92.*

he drew the truss as it still stood. This was because its upper face was against the gable wall and, subjected to the intense heat of the stonework, had been smouldering away long after the fire was out. Not only would the west end of the tiebeam and the two struts of each of the arch-braces have to be renewed, as his drawing showed, but practically half the tiebeam, the entire post of the west side and three quarters of the other post, whilst both of the arch-braces would have to be patched.

*Fig. 96. Truss I as erected, the wrong way round*

There was both delay in obtaining the new timbers and confusion about numbering them and laying out the truss. The plaques attached while the truss was still standing, were necessarily on the "wrong" lower-face of the timbers, but the numbers on them followed the rule of plain digits to the left of the truss centre line, flecked digits to right, as one faced the truss.

In laying it out in bays 6 and 7, exactly as he had the previously repaired trusses, with the tiebeam to the north, Dauncey reproduced Russell's drawing, but as seen from above, as a plan, rather than horizontally, as an elevation. As reared from that position it would have represented the reverse of the drawing, and this in effect is what happened, but by handling each component separately. It was fully repaired and ready for erection on 28 May. It was then noticed that the tiebeam had been repaired at its east end instead of west end. This could easily have been corrected but much had been going on in the meantime.

Truss II and the arcade framing of bay 1 were also being repaired. The sheeted scaffolding of bays 4 and 5 was being taken down and bay 1 scaffolded. The components of truss I and bay 1 had been suitably placed on the floor for raising or hoisting them with block-and-tackle. Clearly it was not simply a matter of turning a tiebeam the other way round, but reorganising the whole of a bay. Nor was it any light task to move massive timbers, especially when standing scaffolding had to be negotiated.

And there was another anomaly. Of the two posts, that to the west should have been wholly renewed, the east post three-quarters renewed. These were also reversed, and of the new components the only ones to have been numbered correctly. All the rest of the numbers as the truss now stood were incorrect and on the wrong face! Hardly surprisingly, we decided to leave everything as it was. Perhaps some future student of ancient timber structures will explain the contradictions, as we did at Frocester. It will hardly occur to others that there is anything to explain. Only the odd strap at the end of the tiebeam may raise an eyebrow. Its purpose was to resist a possible split which looked likely to Dauncey when he was jointing it.

There were no further calamities (yet), but a slight departure from the original was the insertion of a filler piece between the arcade-plate of bay 1 and 2 east side. The mortice of the original had been too badly damaged and the plate of the bay was too secure to raise and drop into the awaiting mortice of the new bay 1 plate. The same operation was necessary later on at post V (west), for the same reason. The west arcade-plate had to be renewed in every bay 1 to 9.

On 21 May truss II was laid out and it was reluctantly decided that post II (west), made famous by its hanging from a pair of pegs after the fire and propped as the first emergency repair, would have to be replaced. The new post when the truss was re-erected was also left hanging. The contractor was instructed several times in the job book to place a pad, hammered in, between its foot and the stone plinth. It was done eventually, but not before the post had begun to twist anticlockwise. It was restrained but too late to prevent some movement. It too remains as evidence of rebuilding, without medieval skill! On 18 June the truss was up and bay 1 complete. Bays 4 and 5 now had to be scaffolded to tackle the in-situ repairs of truss V, scarfing new ends onto the tiebeam, renewing the post-head of the west post, and inserting a very long patch to the outer edge of the east post.

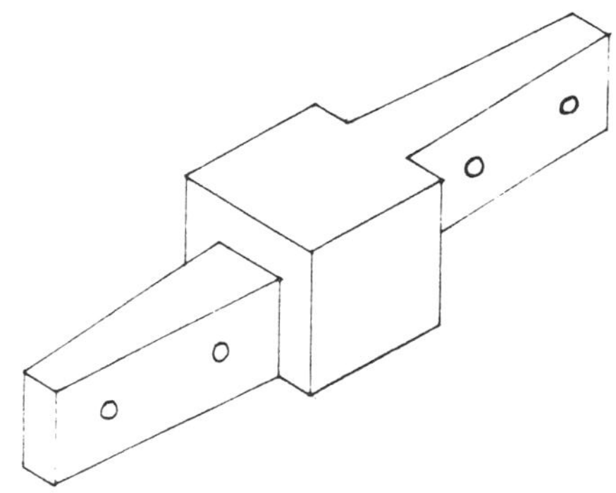

*Fig. 97. Arcade-plate filler-piece as required for arcade-plate I–II*

*Fig. 98. Arcade-plate I–II, condition of joint*

*Fig. 99. Arcade-plate I–II, awaiting filler-piece*

Both of the arcade-plates also had to be renewed, as well as the swept arcade-brace on the west side. After that, it was back (or forward) to truss VI, and it was now July, 1981. Trusses I and II and the intermittent repairs and aisle framing down to truss V had taken three months. From now the erection of trusses, rather than the long slog of repairing, renewing and re-framing them on the floor, took over. Progress would be visible.

Other work during this period had included the resorting of slates by Miles and his men and much discussion about the possible sources for replacements. About twenty-five thousand slates, in all, including the salvaged few, would be required. Miles was sceptical about relying on "new" quarries. Only the still working Guiting quarry could guarantee to fulfil the order. They had already provided the new stone for the corbels and plinths, but their slates were more expensive than Miles was prepared to pay within his estimate. They could make more profit by grinding them down to dust and selling it to Bradstones

*Fig. 100. Arcade-plate I–II joint shown on 94 after renewal*

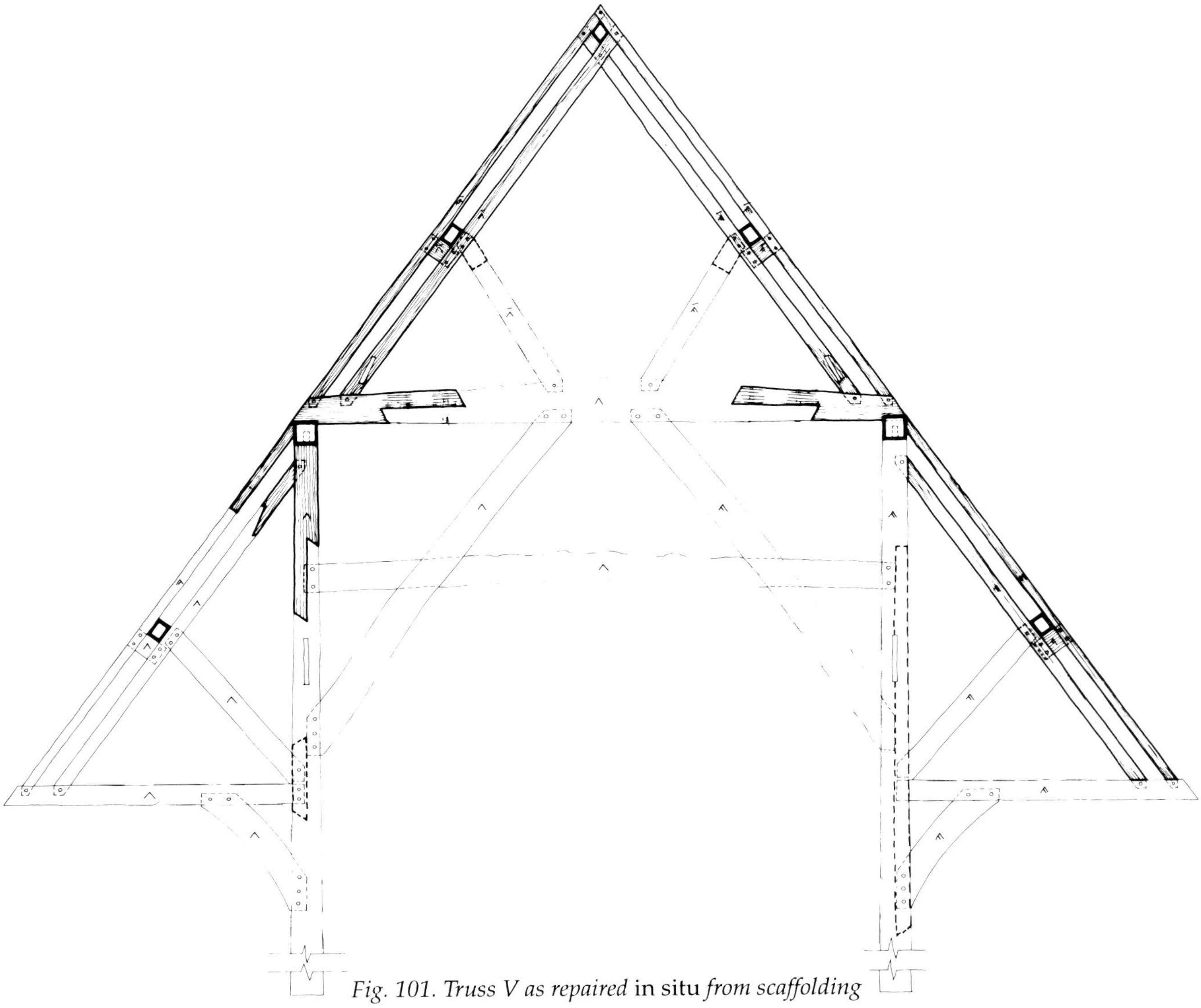

*Fig. 101. Truss V as repaired* in situ *from scaffolding*

for manufacturing their artificial "Cotswold" concrete slates. Miles was also critical of the samples we had obtained from reopened quarries. Clearly we had to leave the choice to him and his own resources – demolished buildings or slates salvaged from old roofs which their owners had reslated in Bradstone!

We had also sought the advice of Miles on fixing the slates. The traditional method of pegging with oak pegs, was expensive and not long-lasting. We had already used aluminium pegs at Middle Littleton and they seemed to have been effective. This and the absence of any reservations on Miles' part was the best assurance we could have.

Replacing the concrete lintols of the two west doorways with oak beams, salvaged by Spicers also from unknown demolished buildings, was also done.

There were still four stages in the re-framing of the barn – first, erecting the trusses of the nave, VI to X, the arcade-framing of bays 5 to 9 and trusses XI, XII and XIII of the north-east porch; secondly, the aisle-framing, already in progress after completion of trusses I to V, consisting of the aisle-ties and their short swept braces, the compound principles and inclined struts, and longitudinally, the purlins; thirdly, the top roof with its compound principals, struts and windbraces between the arcade-plates and purlins, and the diagonally set ridge-piece; and finally, the rafters at both levels, all beginning at the south end of the barn.

If the bulk of the repairs and fabricating of the big timbers was done, there remained most of the framing of the structure, and, of course, getting it up. The coming difficulties and hazards of this were hardly yet realised. It is one thing to frame huge timbers as they lie on the floor. It is quite another when they are swinging from a mobile crane and every joint and peg has to be hammered in from a ladder or swaying scaffolding.

*Figs 102–105. Erection of truss VIII – bringing in the post, placing it on its plinth stone, adjusting the top and bottom, arch-brace fitted to post*

*Figs 106–109. Erection of truss VIII – strainer beam fitted and tiebeam being hoisted and placed onto braces*

Each post was erected first by tying a sling four feet below its head and hoisting it vertically with the "giraffe". Once lowered onto its pad, on the stone plinth, it had to be tied to the top of the scaffolding behind it, close enough to prevent it toppling backwards – or forwards. Gusts of wind could be dangerous. Even worse, were low-flying FI-IIs. Their coming was too swift for anything except to clutch the nearest support and freeze.

The next operation was perhaps the trickiest, when the strainer beam and two arch-braces, half-lapped together on the floor, were hoisted as an open frame, with only the tops of the braces tied to wooden splints. The posts had to be released at the head to lean outwards on ropes until the tenons at each end of the beam and feet of the braces could be engaged. The posts were then drawn in by means of a geared pulley or "Stillson" and the tenons coaxed home, not without the occasional help of a sledge hammer on the outer face of the posts.

When the whole frame was level and secure, the joints could at last be pegged and hammered home. But it was only the scaffolding that held the truss until both sides of the bay-framing (for truss VI this was bay 5) had also been erected. For this operation the two arcade-braces were tenoned into and roped to posts V and VI. The arcade-plate could then be hoisted by the giraffe, now in bay 5, and laid into the tapered mortice of the previous plate and at the same time dropped into the cock-tenons of the post and arcade-braces. Only when that had been completed on both sides was the truss safe. Lastly, the giraffe, brought back to where it started, raised the tiebeam and lowered it onto the dovetails of the plates. At the same time its soffit mortices had to engage the tenons of both braces.

I had previously questioned the wisdom of cutting these mortices until the frame was up. Rightly, however, Dauncey had said that marking them out on the scaffolding, taking the beam down, forming the mortices and then hoisting it back again would delay the whole operation and such niceties as draw-boring would be impossible. When they actually fitted, the truss was complete and secure; it was then time for a celebration.

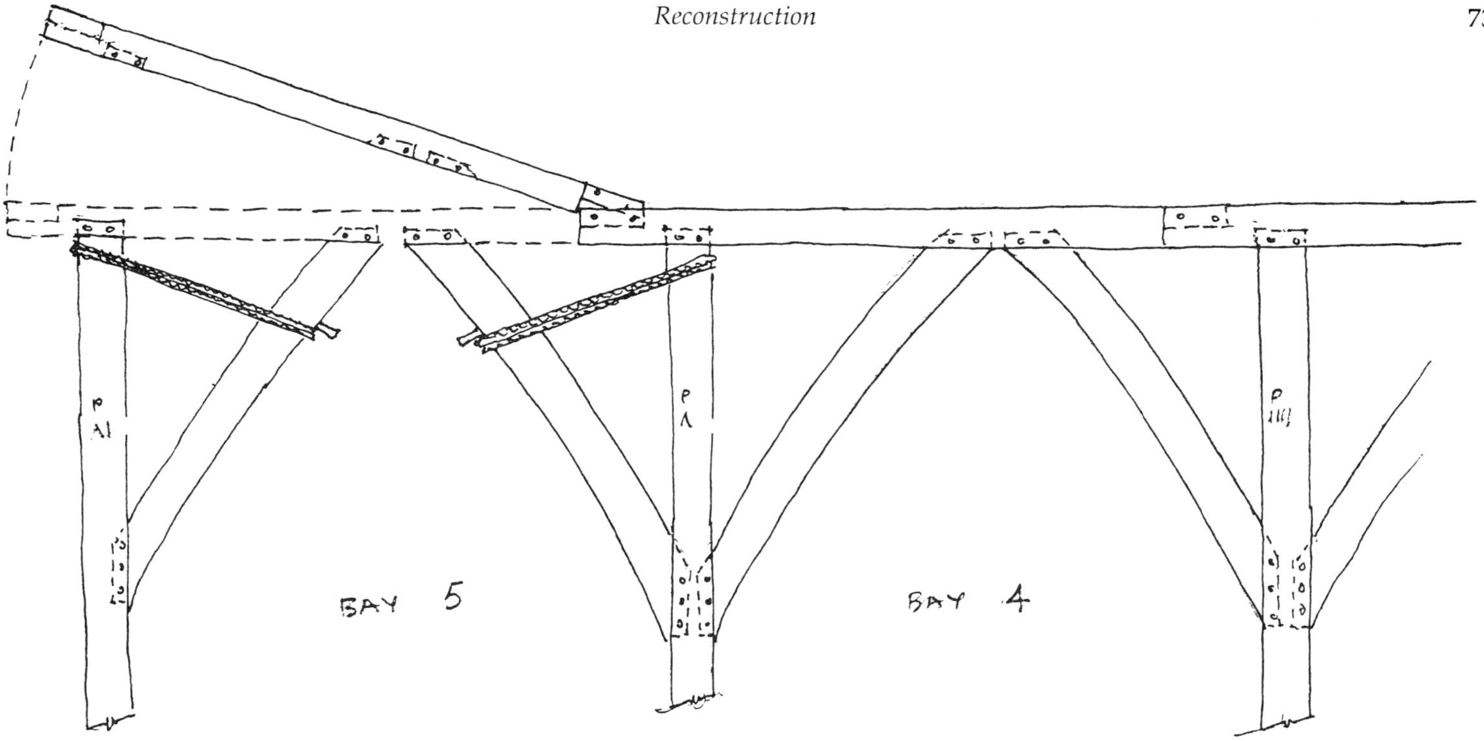


*Fig. 110. Arcading of bays 4 and 5, showing method of tying braces while arcade-plate is lowered*

The erection of truss VI began on 2nd July 1981 and was completed with bay 5 on 16th. On the 25th August truss X was up. Each truss and bay had taken about a fortnight. The appearance of the barn was transformed, particularly at the south end. The aisle framing had reached bay 6 and erection of the top roof was beginning in bay 1. The north east porch trusses were next.

There had been visits by Walter Horn and Abbott Cummings from America, the Essex architect and County Council members concerned with the forthcoming restoration of their dilapidated Coggleshall barn, believed to be the oldest in England, and by Sir Angus Stirling, now National Trust Director and other London trust members. All head office delays were now over and it was decided that even the south porch, untouched by the fire, should also be restored. Dauncey could take a holiday.

*Fig. 111. Main trusses and arcading erected – bays 5–9, seen from west*

*Fig. 112. All main trusses erected – 25 August 1981*

Fig. 113. View from north at arch-brace level, showing varying profiles

Fig. 114. View at strainer-beam level

*Fig. 115. View from north of arcade-plate level*

Then the final disaster! No sooner had Dauncey gone on holiday than it was found that the arcade-plates of bay 9 were nine inches too high for their seating in the north gable. I too was away and Russell optimistically noted in the job book "Mr Charles to investigate plate levels on his return. Instructions will be issued shortly". In fact, instructions were not issued until 15th October. Of course, we should have taken levels long before, or even aimed the plates with string lines from truss VI to the seatings in the gable. But our task in detailing and, still more, the carpenter's in fabricating the joints, would not have been made easier, for the arcade framing, instead of being set out as rectangular, would have to have been out of true. The consequent joints would have been practically impossible either to draw or to make.

We did in fact make large-scale drawings to assess the alterations that would have to be made both to the timbers and to the joints of the arcade-frames of bays 8 and 9. Everything would have to be taken down, and all joints modified. We abandoned this solution and instead considered raising the gable. This too, involving high scaffolding, taking down the big coping stones, obtaining new stone and relaying them would have been not only expensive but, worse, a structural alteration to the stonework of the barn. It too was abandoned.

At last, the obvious question had to be asked: if in the space of six hundred years, the gable wall had subsided nine inches, which is hardly excessive, and the frame had gone down with it, without a suggestion of straining any joint or shearing a single peg, could not the framework again be persuaded to sag a little without visibly altering its form, but in six weeks instead of six centuries?

*Fig. 116. Arcade-plates nine inches too high for original slating in north gable*

*Fig. 117. Measuring-tape for exact discrepancy*

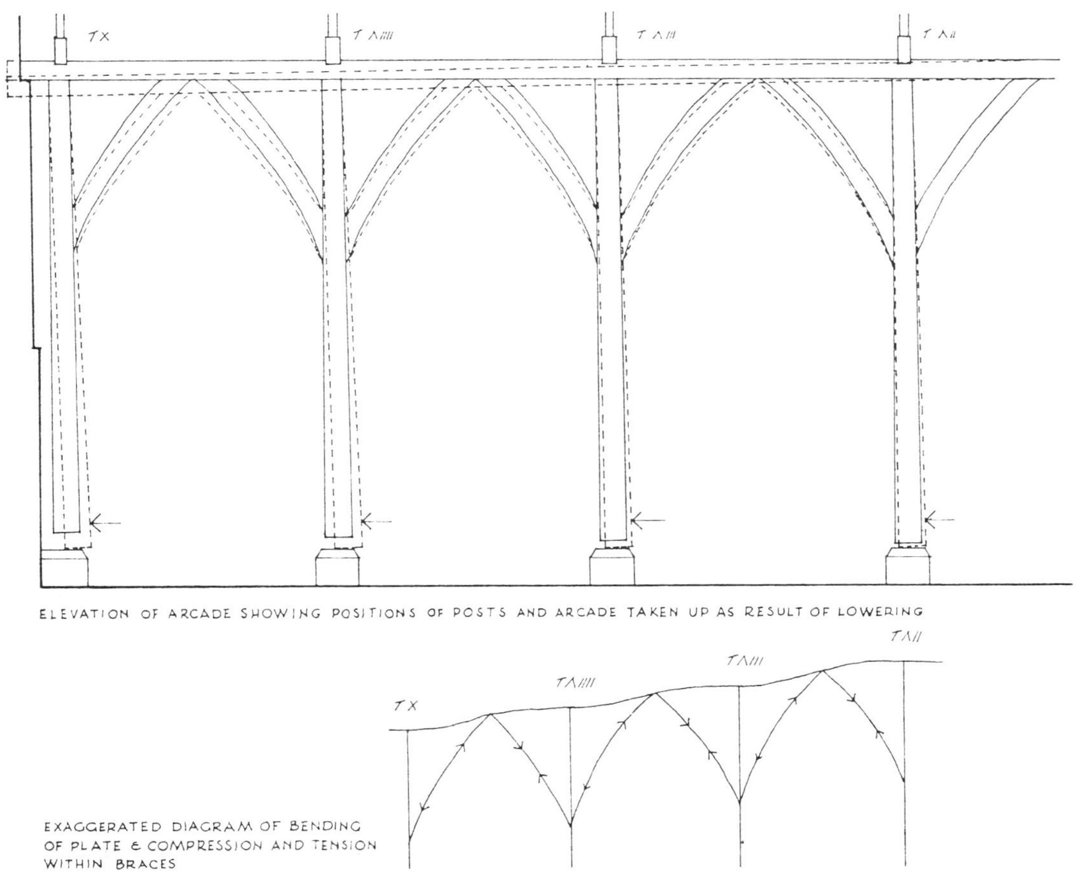


*Fig. 118. Two diagrams illustrating the likely reaction of posts and arcade-plate as a result of reducing the height of the posts – post X by 9 inches down to 2 inches for post VII*

Fig. 119. Post X after cutting

Fig. 120. Post VIIII reduced by 7 inches

Fig. 121. Post VIII reduced by 5 inches

Fig. 122. Post VII reduced by 2 inches

Fig. 123. Job completed 19 November 1981 – from left to right, Messrs Simcock, Chambers and Dauncey

*Fig. 124. Arcade-plate lowered*

Because of the advancing aisle-framing and top roof, Spicers understandably were impatient. On 15th October we adopted this method, reducing the height of the posts progressively, and leaving the joints to look after themselves. The job book reads: "No truss should be lowered more than 1" at any one time. Folding wedges should be placed under the post feet to allow controlled adjustments. Eventually, it is hoped, the arcade-plate will form a natural curve from truss VI down to truss X. At the same time the posts must be kept vertical by inserting acroprops horizontally between the posts. A record of the movement should be kept".

On 3rd November the posts of truss X were sawn off to reduce their height by nine inches. On 12th November the truss had dropped 1¼", so that the arcade-framing and plates cantilevered from truss VIIII were hanging 7¾" above the plinths. On 19th November, having issued detailed instructions for lowering each truss back to VI and an estimate of the amount of movement expected, we cut all the posts from truss VIIII in diminishing amounts down to truss VII in a single afternoon, and let each of them hang, only with blocks under them in case of collapse. In fact, it was adjustment of the acroprops that determined the amount of decline. By tightening them the truss in advance of the fixed truss could be heightened, even back to its original height. But the amount of kick when they were slackened varied with each truss. Thus adjustment was easy, and each day Dauncey would check the tightness of the props and the amount of settlement achieved by each truss. On 26th November, the job book reports "all trusses have now settled down on the plinth pads. Daylight can be seen only under half of the post VII. TX is standing without any strutting and the feet have not kicked at all". All the acroprops were removed before Christmas. Only the principal rafters of TX had to be adjusted later to align with the gable-ridge.

*Fig. 125. Ridge adjusted to new level*

For the rest, raftering out had begun at bay 1 and Miles was battening and slating, with none of the preparation we had witnessed of other roofers. Laying out in rows the different sizes of slates seemed hardly to be necessary, as Miles and his men were able to summon the right slate from haphazard heaps of his roof scaffolding, not only as regards size, but its ability to lie flat and firm on the course below without any kind of bedding. Roofing, nonetheless, was a lengthy process. The eaves had to be repaired with new stone, as already noted, many gable coping stones renewed and all of them rebedded. The wall-plates over the new lintols of the west doors also had to be designed afresh to preserve the correct eaves level and overhang across the openings.

The rest is told by photographs, including the restoration of the south-east porch.

*Fig. 126. Upper roof started at south end, November 1981*

*Fig. 127. Raftering bays 1, 2 and 3. April 1982*

*Fig. 128. Rafter housing in arcade plate*

*Fig. 129. Slates as laid out on west side of barn*

*Raftering as far as north-east porch*

*Figs 130 & 131. Interior of re-raftered barn – June 1982*

*Fig. 132. Completed framing of roof, looking south from truss VII, note strainers intermediately in each bay, probable insertions*

*Fig. 133. View from south-east – north porch rafters, south porch still with original roof*

Fig. 134. South-east porch, re-roofing

Fig. 135. View from south

*Fig. 136. Reeve's offices over porch – restored*

*Fig. 137. Reeve's office looking into barn*

Fig. 138. Exterior of south-east porch restored

*Fig. 139. View from porch and gallery, new west doors*

*Fig. 140. Completed barn, bays 1, 2 and 3 – note porch gallery*

*Fig. 141. Bays 3–9, looking north*

Fig. 142. North gable wall from bay 6 – note oak panel

*Fig. 143. Oak panel on north wall with names of the workers who were there throughout the contract*

*Fig. 144. Completed barn from north-west – note west doors restored even though recently inserted – probably 1951*

## *Summary*

*Fig. 145. 1980*

Monday 18th April 1983 marked completion of the rebuilding of Bredon Barn. Spicers laid on refreshments and a slide-show. The slides, projected onto a giant screen inside the barn, beginning with the fire brought back vivid memories – almost the smell of the smoke – to those who had witnessed the blaze, three years ago to the day.

In the rebuilding, two hundred-and-ninety-eight timbers, excluding rafters had been replaced or repaired. Most of them had to be jointed at both ends; every joint exactly cut by hand and draw-bored and pegged. Their number is uncountable. Only the carpenters proved that they had little to lose in comparison with their medieval forbears.

Mistakes may be found by the more observant and knowledgeable, but there are few ancient buildings of such size and complexity that are perfect. And already it may be seen that the still far too few visitors to the barn find the greatest difficulty in distinguishing the new from the old.

All that is now needed is that the National Trust should acquire the manor house and its grounds, demolish the wall (possibly building another to isolate the shoddy farm buildings on the west side) and restore the barn to its historic landscape and original barton. That its original buildings, as the manor house itself, have gone or been radically altered is of no consequence. The barn will be seen again as first designed, and with the house as a hotel, there will be no shortage of visitors. It could also be safely left open everyday.

## *Accounts*

The final cost of reconstructing Bredon barn including the south-east porch and loft, was £226,849.36, about £24,000 less than the provisional contract signed with Spicers in 1980.

*Fig. 146. Bredon manor house from roof of barn. The original barton was to the right (south) of long range shown (foreground right)*

# NOTES

## Introduction

[1] R Marshall, *William Morris and his Earthly Paradises* (George Brazillier, Inc, New York, 1981) p 227

[2] W Horn and E Born, *The Barns of the Abbey of Beaulieu and its Grange of Great Coxwell and 'Beaulieu St. Leonards'* (University of California Press, 1965). I am grateful for permission to use information and reproduce drawings from this book.

[3] C Hewett, *The Development of Carpentry 1200–1700* (David and Charles, 1969)

## Chapter 1

[1] F Charles and W Horn, *The Cruck-built Barn of Leigh Court, Worcestershire* (Journal of Society of Architectural Historians, 1973 – America)

[2] W Horn and F Charles, *The Cruck-built Barn of Middle Littleton, Worcestershire* (Journal of Society of Architectural Historians, 1966 – America)

[3] C Bond, *The Estates of Evesham Abbey, a preliminary survey of the topography* (Vale of Evesham Historical Society Research Papers 1973)

[4] W Horn and F Charles, *The Cruck-built Barn of Frocester in Gloucestershire* (Journal of Society of Architectural Historians, 1984 – America)

[5] S Rigold, *The Medieval Tithe Barn, Bradford-on-Avon, Wiltshire* (Department of the Environment, 1977). I am grateful to Dr Beric Morley for supplementing both the information of this booklet and the brief survey of the barn by W Horn and myself in 1976. We had intended to add this barn to our list for publication.

## Chapter 2

[1] G Webster and B Hobley, "Aerial reconnaissance over the Warwickshire Avon", in *Archaeological Journal* CXXI (1964) pp 1–22, especially pp 12–13; TC Hencken, "The excavation of the iron age camp on Bredon Hill, Gloucestershire, 1935–1937", in *Archaeological Journal* 95 (1938) pp 1–111

[2] HPR Finberg, *The early charters of the West Midlands* (Leicester, 1961) pp 38, 39, 40–41, 90, 94–95, 102

[3] C Dyer, *Lords and peasants in a changing society. The estates of the bishopric of Worcester 680–1540* (Cambridge, 1980) pp 9–16

4 *Valor ecclesiasticus* (Record Commission, 1810–34) iii, p 267

5 Hereford and Worcester County Record Office (HWCRO), ref 721:091, BA 2358/2, no 34 a; ref 705:139, BA 8397/47 S143/44

6 HWCRO, ref 705:139 BA 8397/47 S143/44

7 Dyer, *Lords and peasants*, p 59

8 TR Nash, *Collections for a history of Worcestershire* (London, 1781), i, pp 134–5

9 Information from Dr DM Barratt

10 A letter from Dr JM Fletcher, 6th May 1982, dated the latest ring on a sample of timber to 1326, and argued that at least 20 years should be added to allow for the missing sapwood

11 HWCRO, ref 009:1, BA 2636/191 92625 1/12; M Hollings (ed), *The Red Book of Worcester* (Worcestershire Historical Society, 1934–50), i, pp 91–111

12 Dyer, *Lords and peasants*, pp 52–55, 79–81

13 Ibid, p 82

14 I Kershaw, *Bolton Priory, The Economy of a Northern Monastery* (Oxford, 1973), pp 18–24

15 Dyer, *Lords and peasants*, p 379

16 Ibid, p 82

17 The calculations of the volume of corn in sheaf that could have been kept in the barn is based on figures in D Oschinsky (ed), *Walter of Henley* (Oxford, 1971) p 475

18 For expenditure on a large barn, JZ Titow, *English rural society 1200–1350* (London, 1969) pp 203–204

19 This information comes from 5 accounts from the period 1375–96: HWCRO, ref 009:1, BA 2636/157 91012 1/8, 158 92014, 92015, 92017, 92020

20 RM Haines (ed), *A calendar of the register of Wolstan de Bransford, Bishop of Worcester 1339–49* (Worcestershire Historical Society, 1966) pp 113–28

21 HWCRO, ref 009:1 BA 2636/158 92015, 92020

22 RH Hilton, *The English peasantry in the later middle ages* (Oxford, 1975) p 89

23 PDA Harvey (ed), *Manorial records of Cuxham, Oxfordshire, Circa 1200–1359* (Oxfordshire Record Society, 50, 1976) pp 44–45

24 HWCRO, ref 009:1, BA 2636/158 92024

25 HWCRO, ref 009:1, BA 2636/175 92479